Maize - Production and Use

Edited by Akbar Hossain

Published in London, United Kingdom

IntechOpen

Supporting open minds since 2005

Maize - Production and Use
http://dx.doi.org/10.5772/intechopen.82912
Edited by Akbar Hossain

Contributors
Elpiniki Skoufogianni, Alexandra Solomou, Georgios Charvalas, Nicholaos Danalatos, Akbar Hossain, M. Shohidul Islam, Mst. Tanjina Islam, Nur Islam, Sarif Sarif Ahmed, Khokan Kumer Sarker, Mahesh Kumar Gathala, John Nedson Ngombe, Kelvin Mulungu, Shikha Bathla, Manpreet Jaidka, Ramanjit Kaur, Md A. Mannan, Adejo Patrick Emmanuel, Msr Krishna, Sagar Maitra, Tanmoy Shankar, Pradipta Banerjee

First published in London, United Kingdom, 2020 by IntechOpen
IntechOpen is the global imprint of INTECHOPEN LIMITED, registered in England and Wales, registration number: 11086078, 7th floor, 10 Lower Thames Street, London, EC3R 6AF, United Kingdom
Printed in Croatia

British Library Cataloguing-in-Publication Data
A catalogue record for this book is available from the British Library

Additional hard and PDF copies can be obtained from orders@intechopen.com

Maize - Production and Use
Edited by Akbar Hossain
p. cm.
Print ISBN 978-1-83880-261-5
Online ISBN 978-1-83880-262-2
eBook (PDF) ISBN 978-1-78984-621-8

For EU product safety concerns:
IN TECH d.o.o., Prolaz Marije Krucifikse Kozulić 3, 51000 Rijeka, Croatia, info@intechopen.com or visit our website at intechopen.com.

Meet the editor

Dr Akbar Hossain is presently working as a senior agronomist at the Bangladesh Wheat and Maize Research Institute (BWMRI), Dinajpur, Bangladesh. Dr Hossain finished his PhD degree on stress physiology of spring wheat in 2013, from the Institute of Ecology and Biological Sciences, Astrakhan State University, Astrakhan-Russia. His research interests and expertise are plant physiology (especially stress physiology), crop management, weed management, weed biology and ecology, conservation agriculture, and climate change impact assessment on field crops through modeling. He has authored more than 142 national and international journal articles. Beside these, he is also serving as a voluntary reviewer/editor in different national and international journals. Further details on Dr Hossain are available in the following link: https://www.researchgate.net/profile/Akbar_Hossain2?ev=hdr_xprf

Contents

Preface

Maize is an important staple cereal after wheat and rice. It is a source of carbohydrate, protein, iron, vitamin B and minerals for many poor people in the world. In some areas, maize is used both as silage and the grains are usually used for food, starch and oil extraction industrially. In developing countries, maize is a major source of income for resource-poor farmers. Due to its multiple uses, the demand for maize is rising day by day globally. Therefore, it is imperative for the improvement of maize productivity to meet the increasing demand. This book, entitled 'Maize - Production and Use', highlights the importance of maize and the improved management approaches for improving the productivity of maize in the era of changing climate.

As the academic editor of the book, I want to thank all the authors for their wonderful contributions, as well as the speed and efficiency in the delivery of their chapters. Special gratitude should be mentioned to all the excellent team from the Editorial Board of IntechOpen, particularly to Ms. Marijana Francetic, Author Service Manager, for their continued support and final compilation of this book.

Akbar Hossain, Ph.D
Senior Agronomist,
Bangladesh Wheat and Maize Research Institute (BWMRI),
Dinajpur, Bangladesh

Section 1

Importance of Maize as Food and Feed Crop

Chapter 1

Maize as Energy Crop

Elpiniki Skoufogianni, Alexandra Solomou,
Georgios Charvalas and Nicholaos Danalatos

Abstract

Maize is the predominant raw material (together with sugar cane) for the production of bioethanol, the most common and widespread biofuel, and at the same time the predominant raw material for biogas production, with the highest yields in Europe. The advantage of maize biomass over other energy plants is the fact that biomass occurs after harvesting the seed and does not require the use of a different area for its development. The main drawback of the use of maize biomass is the negative effects of removing crop residues on fertility and the physical properties of the soil. Bioethanol's share of global biofuel production is over 94%, as many countries are replacing a portion of their fossil fuels with biofuels, according to international regulations. The choice of crops used as feedstock for the production of bioethanol is strongly associated with local climatic factors. About 60% of world bioethanol production is made with cane raw material in the Central and South American countries, with Brazil leading, while the remaining 40% from other crops with North America producing bioethanol almost exclusively from maize, and the EU uses as raw material raw starch (cereals and maize) as well as crops such as sugar beet and sweet sorghum.

Keywords: maize, bioethanol, biodiesel, bioenergy

1. Introduction

As a result of anthropogenic activities, billions of tons of carbon dioxide deriving mainly from the burning of minerals fuels (oil, coal, and natural gas) as well as other gases, such as methane and nitrous oxide, are annually released into the atmosphere, thus changing the composition of the gases that have remained stable for tens of thousands of years [1, 2]. This overturning is expected to change drastically the climate in the next decades. Its dioxide coal is responsible for 50% of the atmosphere's overheating [3, 4].

Despite the environmental burden, the shifting to alternative forms of energy has begun from the oil crisis in the 1970s and the sudden rise in oil prices. This has led to the first boost for the development of renewable energy sources. In addition to food production, many governments supported the development of new cultivated plants for energy production [5, 6].

However, the fall in oil prices in the 1990s tempered the markets, resulting in hindering green energy development and limiting them to small ones.

Nevertheless in our day and age, global energy requirements have increased sharply due to the rapid increase in both the population and the technology. Therefore, alternative forms of energy are imperative. Research has shown that by

2030, the world's population will have grown from 6 to 8bn (33%) and the demand for energy will increase by 50% [7, 8].

Hopefully, there are many possible alternatives to fossil fuels, especially for heat and power generation. In recent years, we have seen a strong desire for some nations to reduce their confidence in fossil fuels and turn to new forms of energy. Three new markets have emerged for energy crop plants:

- bioenergy;
- biofuels; and
- biorenewable materials [9, 10].

Energy crops are either cultivated or native species, traditional or new, which produce biomass as the main product that can be used for various energy purposes [11]. The biomass produced can be used for combustion or cogeneration for coal, electricity and heating as raw material for thermochemical processes such as pyrolysis and gasification for the production of methanol, biogas and pyrolytic oils and for biochemical processes (for example, fermentation) for the production of ethanol or methane [12, 13].

Their main advantage is that their stable production can ensure a large-scale long-term raw material supply with uniform qualitative characteristics in liquid biofuels and energy plants.

Traditional crops whose final product is used to produce energy and biofuels are also considered as energy crops, such as wheat, barley, maize, sugar beet, sunflower, etc.

"New" energy crops are species with high biomass productivity, per unit of land and divided into two categories which are agricultural and forestry. Agricultural energy crops are further distinguished in annual and multiannual years.

Biofuel compared with fossil fuels is considered to be more effective. For example with coal, oil and natural gas to produce 1 MJ of electricity; non-renewable energy consumption is projected to be between 1.7 and 4.2 MJ; biomass values range from 0.1 to 0.4 MJ. In the case of thermal energy, prices are 1.1 and 1.5 for fossil fuels and only 0.01–0.15 for biomass. Although the energy is considered to be CO_2 neutral, in fact there is actually a burden on greenhouse gas emissions due to the process of cultivation and harvesting. However, this charge does not exceed the total emissions of fossil fuels which results in being up to 90% reduced [14].

The amount of land devoted to the cultivation of energy crops for biomass fuels is estimated to account for only 0.5–1.7% of the available agricultural land. Although there are still strong concerns about the production of plants for energy and not for the classic crops purposes such as fee [15], human food production [16] and other related issues [17], there is no doubt that plant biomass is of paramount importance in this field of renewable energy sources, particularly, in the production of biogas and biofuels, through well-designed and organized development programs [15, 18].

2. Energy crops

Energy crops include plants intended for energy production. One of their main strengths is stable production, which can ensure a large-scale, long-term raw material supply. In particular, new crops have significantly higher yields per unit area than conventional ones. Energy plants produce different types of biomass as main products, which can be used for various energy purposes [19].

For the production of liquid biofuels, the energy crops that can be grown are: sunflower and soybeans. For the production of solid biofuels, plants such as cardoon, eucalyptus, canary grass and switchgrass can be used. Finally, sunflower, maize and others can be used to produce biofuel gas [20, 21].

2.1 Advantages of energy crops

There are many potential benefits using energy crops such as increased economic rural development, energy security and environmental benefits [22]. Rural economic development, a compulsory reason for producing energy from crops is the development of a new and profitable crop market taken into account that in recent years, crop prices have been extremely low, which means low profits.

Energy crops can be planted in degraded cultivated land, pastures and land which is currently used for traditional crops. There are 392 million acres of land potentially eligible for energy crops in the United States.

Using energy crops to produce transport fuels could increase our energy security. Currently, the US is importing over 50% of the oil used for transport fuels and is estimating that imports could increase to 75%. Dependence on foreign imports has significant economic social costs [23].

The environmental benefits of using energy crops include water and soil improvement. By reducing the use of herbicides and pesticides that reduces the chances of water pollution and other environmental problems are also reduced due to non-point pollution. Compared to traditional crops, energy crops have increased soil stability, reduced surface water run-off and reduced nutrient and sediment transport [23].

Reducing the emissions of energy crops against fossil fuels for power generation unlike fossil fuels, plants grown for energy crops absorb the amount of carbon dioxide released during combustion, unlike fossil fuels.

Another advantage of using biomass is the avoidance of atmospheric pollution with sulfur dioxide (SO_2) produced during combustion of fossil fuels which contributes to the phenomenon of "acid rain." The sulfur content of biomass is practically negligible.

Finally, energy crops can ensure employment and the retention of rural populations in the border and other agricultural areas, thus contributing biomass to the regional development of the country [24].

2.1.1 Zea mays L.

Maize (*Zea mays* L.) is a member of the *Poaceae* family. It originates from the American continent where thousands of ancient cultures, such as the Indians, the Magyars and the Aztecs, used to grow it. Today, it is one of the most popular cultivations around the world, such as the United States, China, India and Brazil and produces the largest quantities. Maize is a monocotyledon annual plant wind pollinated, both self and cross pollinated.

Since the sixteenth century, its cultivation has spread to all tropical, semi-tropical and many temperate regions worldwide. It is a crop mainly used for human and animal nutrition [25], but in recent decades, the production of biofuels from maize has redefined the purpose of its cultivation. Today, the contribution of maize to biofuels and especially to bioethanol has increased at levels equal to or higher than all energy plants [26–28]. Nitrogen and ash concentrations as well as lignocellulose are two very important factors that define the quality of the raw material in ethanol. These characteristics, in most cases, are based on climatic conditions as well as on the genome of the plant [29].

2.1.1.1 *Maize production*

The main root system of maize is rich and can reach a depth of 2.5 m, although its main bulk grows in the first 60 cm of soil.

The pH range for ideal yields is 6–6.5 while a range of 5.8–7 is generally shown, and there are reports that mention an even greater range of 5–8. Generally, attempts have been made to create varieties that adapt to high or low pH in acidic pH, expecting only 35% of ideal yields and being defined as an optimum pH of 6.8 [30].

The water requirements of maize range from 744 to 901 mm. The irrigation frequency affects the yield of corn seed as they propose an irrigation program where a dose of 15% of the water capacity of the soil will be applied irrigation every 9 days [31, 32].

Increased salinity results in reduced plant leaves, decreased green weight, fresh weight, shorter shoots and root lengthening. However, varieties that are ideally adapted to conditions of high salinity have been developed, as they have particular durability [33, 34]. Still hybrids with respect to pure maize rows show greater tolerance to salts [35].

Corn seed germination may be affected even slightly from 28°C or above as the activity of certain protein-producing enzymes is inhibited by this critical temperature and then [36]. When the temperature increases (in the range of 13–38°C), there is a similar increase in leaf growth rate and photosynthesis rate. Also it was found an increase in photosynthesis rate by increasing the temperature (study range, 13–28°C).

The nutrition of the cobbler in continuously cultivated soil suggests 17–23 kg of nitrogen per hectare, while when there is an increase in organic matter, the addition may be twice as low. For high yields, it is necessary to add potassium as a mature crop of maize which may contain up to 30 kg of potassium per hectare in its plant parts. An experiment in Brazil showed that nitrogen application increased the productivity of grains and dry matter, the calorific power, and the potential for energy generation from maize. Maximum grain yield was obtained with an application of 226 kg ha^{-1} N, resulting in 13.647 kg ha^{-1} of grain yield and 10.968 kg ha^{-1} of total biomass. This biomass presents an energy potential of 11.050 kWh ha^{-1}. Taking the use of only husks and cobs into consideration, it is possible to generate 2712 kWh ha^{-1} of bioenergy [37].

Like energy crops, maize is mainly used for two reasons: (i) for the starchy raw material contained in seeds and the material from which bioethanol is mainly produced [38, 39] and (ii) for the biomass (crop residues) resulting from the removal of the seeds and consisting of leaves, stems and a cone of the blade. Biomass can be used for combustion or production of second-generation bioethanol [27, 40, 41].

The appropriate time of harvesting is when the moisture content of the seeds is between 20 and 30% [42]. Late maturation and flowering of maize cause a greater accumulation of lumps with reduced grain yields and a reduced number of cores per plant.

Maize requires more nitrogen and pesticides than many other crops, thus affecting its energy balance. Increasing the energy potential with ethanol from maize is significantly less than with sugar cane [43].

The choice of varieties with a dry matter content of 30–32% is very important for harvesting date to facilitate the process. Based on the system, FAO maize needs about 45 units of heat to form a new real leaf and about 300 units of heat to fully populate the plant. Early varieties (FAO 150–160) require about 2100 heat units, late (FAO 180–210) approximately 2400 units, while biogas crude maize hybrids (FAO 240–260) require a longer period of 2800–3000 heat units.

3. Biofuels

The use of corn-based biofuels was first introduced into the US as a food additive, but ethanol-maize production increased drastically when conventional fuel prices doubled between 2004 and 2007. Biofuels and rising food prices have contributed to the accumulation of wealth in the agricultural sector, thus increasing the income of farmers, potential value to agricultural land and shifting the relative allocation of resources to the agricultural sector in relation to the rest of the economy [44].

The use of biofuels in the transport sector has become very timely recent years.

In **Tables 1** and **2** below, we can see the liquid biofuel production globally and in each continent separately, up to 2017.

3.1 Biodiesel

The European Commission has adopted the Biofuels Directive in 2009, which requires biofuels to contribute 10% of all transport fuels by 2020 [46, 47].

The two main substitutes for conventional fuels are biodiesel and bioethanol. Biodiesel is used in diesel-powered vehicles, while bioethanol is used in gasoline-powered vehicles. The European Union is the major biodiesel producer. USA, Brazil, Argentina, Indonesia and Thailand along with the EU together produce 85% of all biodiesel worldwide. In 2016, 32.6 billion liters of biodiesel were produced globally. Global biodiesel production is expected to reach 39 billion liters by 2024, corresponding to a 27% increase from 2016. It is important to point out that the cost of biodiesel from the first generation biodiesel feedstock is currently 30% higher than of petroleum-based diesel [48]. Furthermore, it is estimated that 60–80% of the biodiesel production cost stems from the cost of raw materials. All this makes use of low-cost second generation biodiesel feedstock which is very attractive alternative [49].

3.1.1 Production

Europe is the world's largest biodiesel producer (**Figure 1**). Total European production in 2016 is estimated at over 1.5 million tons, with Germany and France

	2000	2005	2010	2015	2016	2017
Total	15.9	34.1	94.4	125	132	143
Bioethanol	12.2	24.5	60.5	82.0	85.6	—
biodiesel	0.78	3.42	18.9	28.9	32,6	—
Other biofuels	2.97	6.16	15.0	14.6	13.6	—

Table 1.
Liquid biofuel production globally (all values in billion liters) [45].

	Africa	Americas	Asia	Europe	Oceania
Total	0.07	101	13.9	19.3	0.29
Biogasoline	0.07	72.1	5.95	4.42	0.2
Biodiesel	0.00	12.5	7.48	13.7	0.06
Other Biofu.	0.00	16.0	0.47	1.13	0.00

Table 2.
Liquid biofuel production in continents in 2016 [45].

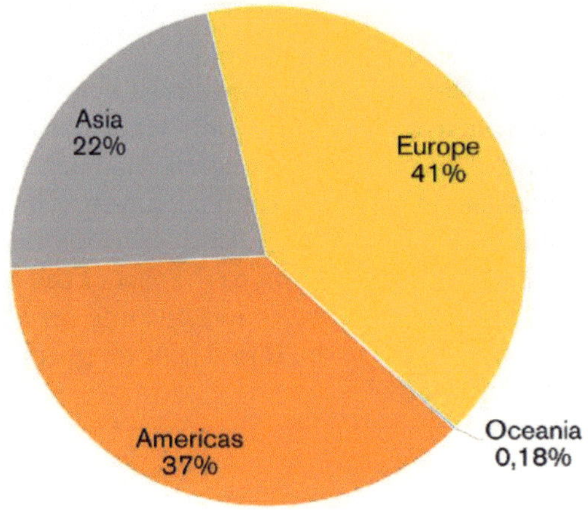

Figure 1.
Liquid biodiesel production in 2016 [45].

being the largest producers within the EU. Italy, the Czech Republic and Austria are also active in biodiesel production [50].

Biodiesel is a generic name for the methyl esters or ethyl esters of organic fatty acids. Biodiesel can be produced from a wide range of seed oils such as oilseed rape, sunflower, soybean and coconut oil.

For example, rapeseed oil is an extremely good substitute for diesel, and it is one of the main oil seeds produced in the European Union. The treatment of plant oil through metallization gives us methyl ester by enabling its ultimate use in diesel vehicles [51].

Seed oils used for biodiesel production come from conventional crops grown by conventional techniques in many parts of Europe. With proper management, crop alternatives may arise as seed oil biodiesel opens a new market for regional economies.

The technology for the production of biodiesel from seed oils has been proven and commercially available for several years. For example, biodiesel is produced from rapeseed by a simple transesterification process, which involves reacting the pulp with small amounts of methanol in the presence of a catalyst. The resultant biodiesel is usually mixed with conventional diesel at the refinery. Biodiesel can also be produced from recycled or used cooking oils, and thus provides a useful outlet for disposal of these oils, which otherwise would have to be disposed of in an environmentally acceptable alternative [52].

3.1.2 Environmental performance

The main advantage of using biodiesel as a transport fuel is that it may have a reduction in greenhouse gas emissions compared to conventional oil use. The use of 100% biodiesel (which is rare) can reduce net CO_2 emissions by 40–50%, respectively, 5% reduces CU2 by 2–2.5% [51].

These calculations are based on a comprehensive life cycle analysis of biodiesel – covering crops, biodiesel production and biodiesel use in the vehicle. In theory, biodiesel can be considered free of carbon, since the carbon emitted during combustion is initially blocked during the growth phase of the cultivated plant. In practice, however, the reduction in emissions from biodiesel from energy crops is lower, because growing and growing plants requires the use of conventional fuels. The use of biodiesel contributes to the creation of an alternative for transport fuels in the context of European Union policy and national climate change policies [51, 53].

Biodiesel can reduce emissions and some other pollutants from vehicles, although this depends on the type of vehicle and the fuel specifications. Biodiesel is a new energy source, aiming to reduce crude oil imports and strengthen security of energy supply in Europe. Biodiesel is easily biodegradable and safe, a property that gives it an advantage for specific uses, such as fuel for boats sailing in ecologically sensitive wetlands.

3.2 Bioethanol

3.2.1 Production

At present, Brazil and the United States (which holds 44% of world production and covered 1.2% of demand for automotive fuel producing 12 billion liters of ethanol) are the largest bioethanol producers of transport fuel worldwide (**Figure 2**), using cane and corn as feedstock, respectively. In Europe, bioethanol is mainly produced from sugar beet and wheat. Spain, Poland and France dominate the bioethanol sector in Europe with a total production of 500,000 tons in 2004. Sweden, Austria and Germany are also active in the production of bioethanol. Production in 2015, after continuing increases, amounted to 58 billion liters. The raw material for bioethanol production is common products from agricultural crops that grow using conventional cultivation techniques in different parts of Europe. Bioethanol production from agricultural crops can be a useful new market for regional economies and help regional development. Bioethanol is prepared by fermenting sugars, starch or cellulose using yeast [54]. The choice of feedstock depends on factors related to cost, technology and economics. Technologies for the production of bioethanol from agricultural products containing sugars and starch are commercially available [55].

Cellulosic materials such as agricultural and forest residues, as well as sorted household waste, are considered as future sources of raw material. However, these materials need to be hydrolyzed before fermentation, using a more complex process than the cereal equivalent. In the long run, cellulosic materials will be considered a potential source of sugars for ethanol production and their use can further reduce CO_2 emissions.

Ethanol production is made from corn grain through two different processes: dry or wet milling. The main difference between the two is the grain processing method. In dry milling, which is the most common procedure, the dried grain is milled into a meal, which is then heated in water to liquefy the starch. Then introduce an enzyme to hydrolyze the starch into sugar, and then is added to ferment the sugar into ethanol and CO_2 [56, 57]. The resulting CO_2 can be used for the production of

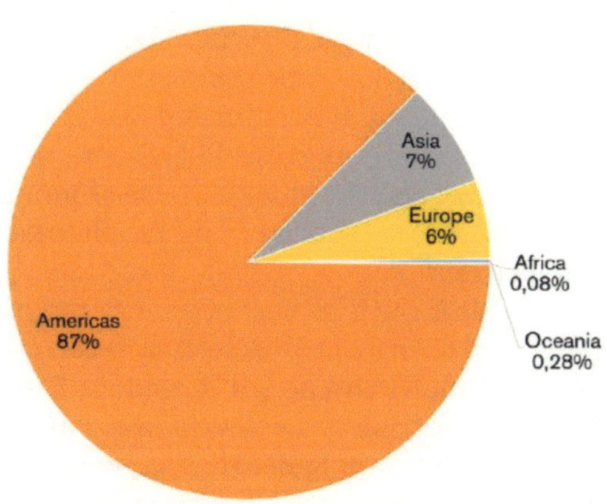

Figure 2.
Liquid bioethanol production in 2016 [45].

carbonated beverages and dry ice and starch-off cereal residues can be marketed for animal feeding (DDGS). During wet milling, the plants, oil (germination) and protein content are separated from the starch (endosperm) in aqueous medium before starch hydrolysis and fermentation begin. With either dry or wet grinding, maize remains a low-cost source of starch that can easily be converted into sugar, fermented and distilled [58].

The choice of crops used as raw material for the production of bioethanol is closely linked to local climatological factors. About 60% of world bioethanol production is produced from sugar cane in the Central and South American countries, with Brazil on the leaf and the remaining 40% from other crops [59], with North America producing bioethanol almost exclusively from maize and the EU uses raw starch (cereals and maize) as well as crops such as sugar beet and sweets. The share of bioethanol in world biofuel production is over 94% with many countries replacing fossil fuels with biofuels [60, 61].

3.2.2 Environmental benefits

The main advantage of bioethanol is that its use results in a significant reduction in greenhouse gas emissions. The use of 100% bioethanol results in a 50–60% reduction compared to conventional fuels. Benefits resulting from the use of blends are obviously smaller [47].

Regarding biodiesel, the benefits of climate change will depend on the raw material to be used to produce bioethanol. GHG (greenhouse gas) emission reductions of 50–60% arise if bioethanol is produced from sugar beet and wheat. If cellulosic materials are used, the net reduction may be greater – perhaps up to 75–80%. This is because less energy is needed for the cultivation of such plants, as well as the fact that during the production phase, energy efficient processes are also used, which also allow the use of renewable energy sources [47].

It is important to understand that bioethanol production is in itself an energy-intensive process and requires significant amounts of energy produced from conventional fuels. However, it is clear that the use of bioethanol can help to achieve the objectives of legislation to prevent climate change. The use of bioethanol can also reduce emissions of other pollutants from vehicles, although this reduction depends on vehicle type and fuel specifications [55, 62].

3.2.3 Disadvantages

There are many concerns about energy crops and bioenergy due to the land and resources needed to produce biofuels. Bioethanol demand in the EU in 2010 amounted to 12.7 billion liters, with domestic production capacity of only 2 billion liters per year [63], so to meet demand it is estimated that it would be about 13% of the total arable land to be used for energy crops [64]. There are serious reactions to the increase in the price of maize and the change of use of limited resources such as cultivated land and water reserves. The use of lignocellulosic corn biomass is an alternative source of biofuels [65].

A major problem in biofuels is the high cost of energy you need to make biomass actively converted [49]. This problem can be solved by research in order to improve the biomass conversion technologies and how it is produced. An important step in the technological field in this direction is the development of second-generation bioethanol production technology from lignocellulosic raw materials, allowing even greater flexibility in the choice of raw materials, releasing much of the arable land from energy production [66].

Finally, a great deal of concern is also given to the biofuels' performance ratio and more specifically to maize from the surveys that have been done to show that the energy efficiency index is positive and can reach 1.5 with more realistic consensus values is 1.25. The net solar conversion efficiency is very low 0.01% (below our initial estimate of 0.045%) [67, 68].

Although second generation bioethanol production technologies from lignocellulosic biomass are still growing, the contribution of maize biomass to bioenergy production is important. The advantage of maize biomass over other energy plants, such as *Miscanthus* x *giganteus*, *switchgrass* (*Panicum virgatum* L.) and others, is the fact that biomass occurs after harvesting the seed and does not require the use of a different area for its development. The main drawback of the use of maize biomass is the negative effects of removing crop residues on fertility and the physical properties of the soil [69, 70].

Corn ethanol is the third most efficient biofuel that yields 1350 l ethanol per hectare. The average US yield in maize is 8.6 mg grains/hectare. Assuming that 25 kilograms of corn grains produce about 10.6 liters of ethanol (a metric equivalent of 1 pounds yields US $ 2.8), the average grain yield translates to 3650 liters/hectare. According to some estimates, the use of ethanol produced from corn cereals offers a 10–20% reduction in GHG emissions compared to petroleum fuels. Maize seed (stem, bark and pellet residues) has the potential to contribute substantially to the biofuel tank when appropriate conversion technologies are developed to convert cellulosic biomass to biofuels. Residues account for about 50% of the cultivation biomass and are readily available in the maize production areas [71].

Several issues need to be resolved before large-scale maize is used to produce biofuels, for example, biodegradation should be at a relatively close distance (about 80 km). From areas where the site will be harvested, transport costs are reduced. The "window" for harvesting the stover will be rather narrow in most places if not removed from the domain.

However, in order for maize to have a sustainable outlook as an energy plant, it is important that the Net Energy Balance (NEB), in the overall production of biofuels from maize growing, be larger than the unit. The term NEB is defined as the fraction between outflows and inputs of the system. The input is considered to be the sum of the fossil fuels required throughout the biofuel production process and includes inputs during the installation and completion of the crop in the field (fertilizers, use of agricultural machinery, agrochemicals, etc.), transportation and the process of converting the seed or biomass into biofuels and as the output of the total energy of the biofuels produced that eventually end up outside the production system. The energy balance in the production of biofuels from maize is reported in the literature in many larger unit studies [72, 73] but also smaller. These differences in the results of the research are identified in the different biofuel production processes but mainly in environmental factors such as climatic and soil conditions, as well as in the cultivation practices followed and influenced the growth and production of maize cultivation [74] since NEB is mainly determined by crop productivity [75].

3.3 Biogas

Biogas production from energy crops is of increasing importance, as it offers significant environmental benefits such as reducing CO_2 emissions. In addition, it can contribute to raising farmers' incomes. Maize has great potential for biogas production. Biogas has the advantage that it can be used in many sectors, such as car fuel, but also as a source of energy in fixed units. Biogas has greater

	2000	2005	2010	2015	2016
Biogas (Billion m^3)	13.2	23.1	38.7	60.0	60.8
Biogas (EJ)	0.28	0.50	0.84	1.30	1.31

Table 3.
Biogas production globally from 2000 to 2016 [45].

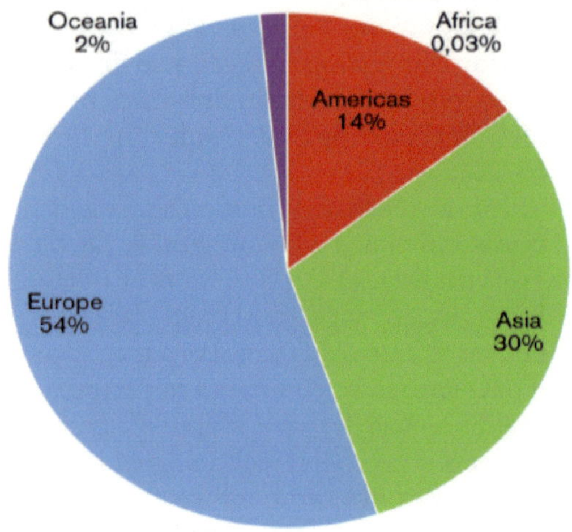

Figure 3.
Biogas production in continents in 2016 [45].

advantages over other biofuels, such as bioethanol for the greater energy that produces, for example, a hectare of corn when converted to bioethanol, giving 20 Gj (Giga Joules). Biogas in the same area gives us nearly three times as much as 55 Gj. Maize, energy beet, rye and grass are crops grown commonly in the central, south-eastern Europe and United Kingdom for energy purposes and mainly for biogas production [74].

Silage maize is digested anaerobically, a conversion process where organic matter of biomass is converted into methane in four phases by bacteria in the absence of oxygen. The end products of the digestion process are biogas and digestate [76, 77].

A major problem we face with maize is its lignocellulos structure which prevents the process of fermenting. Several technologies have begun solving this problem, making maize commercially viable [78, 79]. To help increase the fermentation rate, we cut maize much shorter than a standard loader to increase the surface, which means it will be more accessible to microbes [80].

Recently, lignocellulosic materials have gained more interest as potential candidates for biogas production, but a large-scale implementation has not been widely adopted, mainly because of the complicated structure of the cell walls of lignocellulosic plants, which makes them resistant to hydrolysis by microbial attack. Therefore, the pretreatment of lignocellulosic material is essential step to achieve high process yields [81] (**Table 3**) (**Figure 3**).

4. Conclusions

The rapid development of technology and the constant increase in the number of the world's population combined with the pollution of the environment lead to the need to find new energy resources more friendly and efficient. Energy crops can provide a large amount of energy by exploiting unused agricultural pieces of land or degraded land without burdening environments compared to fossil fuels. Maize is one of the best representatives of energy crops and presents great prospects in the bioethanol sector. Despite the great prospects of energy crops, and in particular maize, we still need research into more efficient use of biomass in cheaper and more economical ways.

Conflict of interest

The authors declare no conflict of interest.

Author details

Elpiniki Skoufogianni[1*†], Alexandra Solomou[2†], Georgios Charvalas[1†] and Nicholaos Danalatos[1†]

1 Department of Agriculture, Crop Production and Rural Environment, University of Thessaly, Volos, Greece

2 Institute of Mediterranean and Forest Ecosystems, Hellenic Agricultural Organization "DEMETER", Athens, Greece

*Address all correspondence to: eskoufog@uth.gr

† These authors contributed equally.

References

[1] Abdus S, Toshikuni N. Impact of human activities on carbon dioxide (CO_2) emissions: A Statistical Analysis. The Environmentalist. 2005;**25**(1):19-30. doi.org/10.1007/s10669-005-3093-4

[2] Hao X, He J, Zhao H, Zhang Q. Investigation of the absorption coefficient of methane in pollution gas measurement. Optik—International Journal for Light and Electron Optics. 2015;**126**(14):1385-1388. DOI: 10.1016/j.ijleo.2015.04.016

[3] Sabine CL, Feely RA. Climate and climate change. Carbon Dioxide. Encyclopedia of Atmospheric Sciences. Eds John Pyle & Fuging Zhang, Elsevier Ltd. 2015. pp. 10-17. DOI: 10.1016/b978-0-12-382225-3.00095

[4] Florides AG, Christodoulides P. Global warming and carbon Dioxide through sciences. Pubmed. Environmental International. 2008;**35**(2):390-401. DOI: 10.1016/j.envint.2008.07.007

[5] Karp A, Halford N. Energy crops: Introduction. Series. Nigel GH, Angela K, editors. Royal Society of Chemistry. RSC Energy and Environment. 2010;**1**:1-12. DOI: 10.1039/9781849732048-00001

[6] Sims REH, Hastings A, Schlamadinger B, Taylor G, and Smith P. Energy crops: Current status and future prospects. Global Change Biology. 2006;**12**(11):2054-2076. DOI: 10.1111/j.1365-2486.2006.01163.x

[7] Glenn J, Warner K. The 21st century population-energy-climate nexus. Elsevier LTD, Energy policy. 2016;**93**:206-212. DOI: 10.1016/j.enpol.2016.02.044

[8] Directive 2009/30/EC of the European Parliament and of the Council of 23 April 2009 amending Directive 98/70/EC as regards the specification of petrol, diesel and gas-oil and introducing a mechanism to monitor and reduce greenhouse gas emissions and amending Council Directive 1999/32/EC as regards the specification of fuel used by inland waterway vessels and repealing Directive 93/12/EEC. OJ, L 140; 2009. pp. 88-113

[9] Röder M, Welfle A. Bioenergy. Managing Global Warming. Eds Trevor Letcher, Elsevier Academic press. 2019. pp. 379-398. DOI: 10.1016/b978-0-12-814104-5.00012

[10] Eichhorn S, Gandini A. Materials from Renewable Resources. MRS Bulletin, Cambridge Core. 2010;**35**(3):187-193. DOI: 10.1557/mrs2010.650

[11] Harvey M, Pilgrim S. The new competition for land: Food, energy, and climate change. Food Policy. 2011;**36**:S40-S51

[12] Ness J, Moghtaderi B. Chapter 1: Biomass and Bioenergy, Coal-Biomass Cofiring Handbook. Cooperative Research Centre for Coal in Sustainable Development. 2007. pp. 1-36. Available from: https://www.researchgate.net/publication/309127353_Chapter_1_Biomass_and_Bioenergy

[13] Voloshin RA, Rodionova MV, Zharmukhamedov SK, Nejat Veziroglu T, Allakhverdiev SI. Review: Biofuel production from plant and algal biomass. International Journal of Hydrogen Energy. 2016;**41**(39):17257-17273. DOI: 10.1016/j.ijhydene.2016.07.084

[14] Ingenito A, Andriani R, Agresta A, Gamma F. A comparative study of combustion between biofuels and fossil fuels. 2012. DOI: 10.2514/6.2012-4054

[15] Cherubini F, Stroemman AH. Production of biofuels and biochemicals from lignocellulosic

biomass: Estimation of maximum theoretical yields and efficiencies using matrix algebra. Energy and Fuels. 2010;**24**(4):2657-2666

[16] Knocke C, Vogt J. Biofuels–challenges & chances: How biofuel development can benefit from advanced process technology. Engineering in Life Sciences. 2009;**9**(2):96-99

[17] Serio MA, Kroo E, Wojtowicz MA. Biomass pyrolysis for distributed energy generation. Preprint Symposium American Chemical Society, Division of Fuel Chemistry. 2003;**48**(2):584-589

[18] Larson ED. A review of life-cycle analysis studies on liquid biofuel systems for the transport sector. Energy for Sustainable Development. 2006;**10**(2):109-126

[19] Jezierska-Thöle A, Rudnicki R, Kluba M. Development of energy crops cultivation for biomass production in Poland. Renewable and Sustainable Energy Reviews. 2016;**62**:534-545. DOI: 10.1016/j.rser.2016.05.024

[20] Claverie M, Demarez V, Duchemin B, Hagolle O, Ducrot D, Marais-Sicre C, et al. Maize and sunflower biomass estimation in Southwest France using high spatial and temporal resolution remote sensing data. Remote Sensing of Environment. 2012;**124**:844-857. DOI: 10.1016/j.rse.2012.04.005

[21] Epplin FM. Cost to produce and deliver switchgrass biomass to an ethanol-conversion facility in the southern plains of the United States. Biomass and Bioenergy. 1996;**11**(6):459-467. DOI: 10.1016/s0961-9534(96)00053-0

[22] Dembris A. Importance of biomass. Energy Sources. 2004;**26**:361-366

[23] Tolbert VR, et al. Soil and water quality aspects of herbaceous and woody energy crop production: Lessons from research-scale comparisons with agricultural crops. In: Conference Proceedings on Bioenergy. Madison; 1998. p. 1272

[24] Western Regional Biomass Energy Program. Environmental Issues. Biomass Digest. 2017;**4**(2):3

[25] Klopfenstein TJ, Erickson GE, Berger LL. Maize is a critically important source of food, feed, energy and forage in the USA. Field Crops Research. 2013;**153**:5-11. DOI: 10.1016/j.fcr.2012.11.006

[26] OECD. Proceedings of Biomass and Agriculture—Sustainability Markets and Policies. Paris: OECD; 2015. ISBN: 92-64-10555-7

[27] Karp A, Shield I. Bio energy from plant and sustainable yield challenge. New Phytologist. 2008;**179**:15-33

[28] Dhugga K. Maize Biomass Yield and Composition for Biofuels. Crop Science. 2006;**47**(6):2211-2227. DOI: 10135.2/cropsci2007.05.0299

[29] Muhaji, Sutjahjo DH. The characteristics of bioethanol fuel made of vegetable raw materials. The Consortium of Asia-Pacific Education Universities (Cape U). IOP Conferences Series: Materials Science and Engineering. 2018;**296**:1-7. DOI: 10.1088/1757-899X/296/1/012019

[30] Greaves JA, Rufener GK II, LeRette RJ, Stoecker MA. High pH tolerant corn and the production thereof. United Stated Patent. 1996

[31] Kara T, Biber C. Irrigation frequencies and corn (*Zea mays* L.) yield relation in northern Turkey. Pakistan Journal of Biological Sciences—PJBS. 2008;**11**(1):123-126

[32] Kranz WL, Irmak S, Van Donk J, CD Yonts, Martin DL. Irrigation Management for Corn. NebGuide.

Lincoln Extension: University of Nebraska; 2008

[33] Faustino FC, Garcia RN, Agtarap ML, Tecson-Mendoza EM, Lips SH. Salt tolerance in corn: Growth responses, ion accumulation, nitrate reductase and PEP-carboxylase activities. Philippine Journal of Crop Science. 2000;**25**(1):17-26

[34] Akram M, Ashraf R Ahmad M, Waraich J Iqbal EA, Mohsin M. Screening for salt tolerance in maize (*Zea mays* L.) hybrids at an early seedling stage. Pakistan Journal of Botany. 2010;**42**(1):141-154

[35] Maiti RK, Kousik SK, González Rodríguez H, Rajkumar D, Vidyasagar P. Salt tolerance of twelve maize hybrids at the seedling stage. Acta Agronomica Hungarica. 2010;**58**:21-29

[36] Riley GJP. Effects of high temperature on the germination of maize (*Zea mays* L.). Planta. 1981;**151**:68-74

[37] Ambrosio R, Pauletti V, Barth G, Povh F. Energy potential of residual maize biomass at different spacings and nitrogen doses. Ciencia e Agrotecnologia. 2017;**41**(6):626-633. DOI: 10.1590/1413-70542017416009017

[38] Morris M, Hill A. Ethanol Opportunities and Questions. Fayetteville, Arkansas: ATTRA–National Sustainable Agriculture Information Service; 2006

[39] Persson T, Garcia A, Paz J, Jones J. Maize ethanol feedstock production and net energy value as affected by climate variability and crop management practices. Gerrit Hoogenboom Agricultural Systems. 2009;**100**:11-21

[40] Sanchez OJ, Cardona CA. Trends in biotechnological production of fuel ethanol from different feedstocks. Bioresource Technology. 2008;**99**(13):5270-5295

[41] Balat M. Production of bioethanol from lignocellulosic materials via the biochemical pathway: A review. Energy Conversion and Management. 2011;**52**(2):858-875

[42] Agronomy corn. Grain Harvesting. University of Wisconsin. 2006. Available from: http://corn.agronomy.wisc.edu

[43] Donato P, Finore I, Poli A, Nicolaus B, Lama L. The production of second generation bioethanol: The biotechnology potential of thermophilic bacteria. Journal of Cleaner Production. DOI: 10.1016/j.jclepro.2019.06.152

[44] Xavier AMRB, Correia MF, Pereira SR, Evtuguin DV. Second-generation bioethanol from eucalypt sulphite spent liquor. Bioresource Technology. 2010;**101**(8):2755-2761

[45] WBA Global Bioenergy Statistics. Summary Report. 2018. Available from: https://worldbioenergy.org/

[46] Rutz D, Janssen R. Biofuel Technology Handbook, WIP Renewable Energies ISBN contract No. EIE/05/022/SI2.420009. Mónchen, Germany; 2008

[47] Pahl G. Biodiesel: Growing a New Energy Economy. Vermont: Chelsea Green Publishing Company; 2005

[48] Rouhany M, Montgomery H. Global Biodiesel Production: The State of the Art and Impact on Climate Change. 2018. doi.- org/10.1007/978-3-030-009854_1

[49] Bušić A, Kundas S, Morzak G, Belskaya H, Mardetko N. Recent Trends in Biodiesel and Biogas Production. Food Technology and Biotechnology. 2018;**56**(2). DOI: 10.17113/ftb.56.02.18.5547

[50] Tyson KS. Biodiesel Technology and Feedstock. NREL. State of Michigan, Senate Ag Preservation Task Force,

Senate Ag Preservation Task Force Report 2003; 1999. p. 1

[51] Biodiesel. 2018. Available from: www.biodiesel.org. [Assessed: 25 June 2019]

[52] Tickell J. From the Fryer to the Fuel Tank: The Complete Guide to Using Vegetable Oil as an Alternative Fuel. Hollywood, CA: Tickell Energy; 2003

[53] Kim SD, Dale BE. Environmental aspects of biofuel derived from no-tilled corn grain: Nonrenewable energy consumption and greenhouse gas emissions. Biomass and Bioenergy. 2005;**28**:475-489

[54] Zabochnicka M, Sławik L. Bioethanol-Production and Utilization, 2010. https://www.researchgate.net/publication/228351087_BioethanolProduction_and_Utilizaton

[55] Dufey A. Biofuels Production, Trade and Sustainable Development: Emerging Issues. Environmental Economics Programme/Sustainable Markets Group; 2006

[56] Prasad RK, Chatterjee S, Mazumder PB, Gupta SK, Sharma S, Vairale MG, Gupta DK. Bioethanol production from waste lignocelluloses: A review on microbial degradation potential. Chemosphere. PMID 31154237. 2019;**231**:588-606. DOI: 10.1016/j.chemosphere.2019.05.142

[57] Sewsynker-Sukai Y, Gueguim Kana EB. Simultaneous saccharification and bioethanol production from corn cobs: Process optimization and kinetic studies. Bioresource Technology. 2018;**262**:32-41. DOI: 10.1016/j.biortech.2018.04.056

[58] Demirbas A. Bioethanol from cellulosic materials: A renewable motor fuel from biomass. Energy Sources. 2005;**27**(4):327-337

[59] Gonzalez-Garcia S, Gasol CM, Gabarrell X, Rieradevall J, Moreira MT Feijoo G. Environmental aspects of ethanol-based fuels from Brassica carinata: A case study of second generation ethanol. Renewable and Sustainable Energy Reviews. 2009;**13**(9):2613-2620

[60] Mussatto SI, Dragone G, Guimaraes PMR, Silva JPA, Carneiro LM, Roberto IC, et al. Technological trends, global market, and challenges of bio-ethanol production. Biotechnology Advances. 2010;**28**(6):1873-1899

[61] David P, Marklein A, Megan A, Karpoff TM, Gillian S, McCormack PR, et al. Biofuel Impacts on World Food Supply: Use of fossil fuel, land and water resources. Energies. 2008;**1**(2):41-78. DOI: 10.3390/en1010041

[62] Zarzyycki A, Polska W. Bioethanol production from sugar beet-European and polish perspective. In: The First TOSSIE Workshop on Technology Improvement Opportunities in the European Sugar Industry; 25-26 January 2007; Ferrara, Italy; 2007

[63] Magar SB, Pelkonen P, Tahvanainen L, Toivonen R, Toppinen A. Growing trade of bioenergy in the EU: Public acceptability, policy harmonization, European standards and certification needs. Biomass and Bioenergy. 2011;**35**(8):3318-3327

[64] Bacovsky D, Mabee W, Worgetter M. How close are second generation biofuels? Biofuels, Bioproducts and Biorefining. 2010;**4**(3):249-252

[65] BP. Statistical Review of World Energy 2010. Bp.com/statistical review

[66] Naylor R, Liska A, Burke M, Falcom WP, Gaskell J, Rozelle S, et al. Ripple effects of crop-based biofuels on global food security and the environment. Environment. 2007;(49):30-43

[67] Perlack RD, Wright LL, Turhollow AF, Graham RL, Stokes BJ, Erbach DC. Biomass as Feedstock for a Bioenergy and Bioproducts Industry: The Technical Feasibility of a Billion-Ton Annual Supply. Oak Ridge, TN: USDOE; 2005

[68] Wright LL, Boundy B, Perlack B, Davis S, Saulsbury B. Biomass Energy Databook. 1st ed. Oak Ridge, TN: USDOE; 2006

[69] Ragauskas AJ, Williams CK, Davison B, Britovsek G, Cairney J, Eckert CA, et al. The path forward for biofuels and biomaterials. Science. 2006;**311**(5760):484-489

[70] Kumar D, Singh V. Bioethanol Production From Corn. Corn. 2019. pp. 615-631. DOI: 10.1016/b978-0-12-811971-6.00022-x

[71] Tilman D, Hill J, Lehman C. Carbon-negative biofuels from low-input high-diversity grassland biomass. Science. 2006;**314**:1598-1600

[72] Patzek TW, Anti SM, Campos R, Ha KW, Lee J, Li B, et al. Ethanol from corn: Clean renewable fuel for the future, or drain on our resources and pockets? Environment, Development and Sustainability. 2005;7:319-336

[73] Hoogenboom G. Contribution of agrometeorology to the simulation of crop production and its applications. Agricultural and Forest Meteorology. 2000;**103**:137-157

[74] Martinov M, Golub M, Viskovic M, Djatkov D, Krstic J. Study of Harvest, Storage and Processing of Corn Stover for its Use as a Fuel and Feedstock for Biofuels in Autonomous Province Vojvodina. NoviSad, Serbia: Faculty of Technical Sciences; 2016. Available from: http://www.psemr.vojvodina.gov.rs/images/Studije/Biomasa/Kukuruzovina_kao_energent_i_sirovina_za_biogorivo.pdf

[75] Lorenz D, Morris D. How Much Energy Does it Take to Make a Gallon of Ethanol? Washington D.C: Institute for Local Self-Reliance (ILSR); 1995

[76] Verstraete W. Biogas. 1st ed. Stichting Leefmilieu v.z.w., Antwerpen VZW – dossier nr. 6. p. 207

[77] Mol APJ. Bounded Biofuels? Sustainability of Global Biogas Developments. Science, Technology and Sustainability. John Wiley & Sons Ltd; 2013. https://doi.org/10.1111/soru.12026

[78] Lizasoain J, Trulea A, Gittinger J, Kral I, Piringer G, Schedl A, et al. Corn stover for biogas production: Effect of steam explosion pretreatment on the gas yields and on the biodeg-radation kinetics of the primary structural compounds. Bioresource Technology. 2017;**244**:949e956. DOI: 10.1016/j.biortech.2017.08.042

[79] Schroyen M, Vervaeren H, Hulle SWH, Van Raes K. Bioresource technology impact of enzymatic pretreatment on corn stover degradation andbiogas production. Bioresource Technology. 2014;**173**:59e66. DOI: 10.1016/j.biortech.2014.09.030

[80] Oslaj M, Mursec B, Vindis P. Biogas production from maize hybrids. Biomass and Bioenergy. 2010;**34**(11):1538-1545. DOI: 10.1016/j.biombioe.2010.04.016

[81] Maryam M, Forgács G, Horváth I. Biogas from Lignocellulosic Materials. Keikhosro K, editors. Springer, Cham. 2015. doi.org/10.1007/978-3-319-14033-9_6

Chapter 2

Nutritive Value

Shikha Bathla, Manpreet Jaidka and Ramanjit Kaur

Abstract

Nature has blessed the human and animal beings with great food diversity in terms of cereal grains to maintain their health status. Among the cereal grains, wheat, rice and maize (*Zea mays*) are the major ones that are considered as stable food across the globe due to their high nutritional significance enriched with abundant amount of macronutrients like starch, fibre, protein and fat along with micronutrients like B-complex vitamins, ß-carotene and essential minerals, i.e. magnesium, zinc, phosphorus, copper, etc. Maize is also considered as low-cost-high-benefit ratio for human beings that help in the prevention of metabolic syndrome due to the presence of different antioxidants like phenols and phytosterols in it. Maize or corn can be consumed only after processing into different food items such as popcorn, flour, tortillas, cornflakes, corn germ oil, etc. Maize products are also used in supplementary nutritional programmes to feed the malnourished children and to improve their health status. However, the quality of maize products depends upon the agronomic practices and climatic conditions.

Keywords: maize, nutritional value, health, quality

1. Brief overview

The cereal grains, wheat, rice and maize (*Zea mays*), are considered as stable food across the globe and contribute to 50–60% of daily human energy requirements. Maize as a third leading cereal grain around the world due to its high yield and nutritive value is also known as the queen of cereal crops. The largest producer of maize is the United States of America (USA) contributing about 35% of the total world maize production. It is also known as the mother grain of Americans and it is the driver of the US economy [1]. Maize is a stable cereal very popular due to its high nutritional significance enriched with abundant amount of macronutrients like starch, fibre, protein and fat along with micronutrients like vitamin B complex, ß-carotene and essential minerals, i.e. magnesium, zinc, phosphorus, copper, etc. Maize also contains a booster of antioxidant that protects from various degenerative diseases. The quality of maize depends upon the agronomic practices and climatic conditions. Maize contains 11% of protein but is deficient in amino acids like tryptophan and lysine. However, new fortified varieties are also being produced in the American region. Maize has to undergo different food processing methods like grinding, alkali processing, boiling, cooking, fermentation, etc. so that it can be used in the development of variety of food items, that is, flour, flakes, popcorn, tortillas, etc. Nutritional quality of maize also depends upon the processing method used for food preparations. Maize like wheat and rice is also used in supplementary nutrition programmes and integrated child development service programmes to feed malnourished children [2]. Maize kernel is an edible and nutritive part of the plant. Maize is a major cereal crop for both livestock feed and human nutrition.

2. Food processing techniques for maize

There are three major processes utilised in the production of maize for food usage as discussed below.

2.1 Dry milling

Grinding of the whole grain stone or roller mill to produce flour or meal is a simple method used around the world when the ground products are to be consumed shortly after processing. The stability of such products is limited owing to the presence of crushed germ in the flour. Oil from broken germ cells is easily oxidised to produce rancid odour and flavour. The large as well as small grits are used in the production of cornflakes and breakfast cereals. Dry milling germ can be pressed or solvent extracted to recover the valuable oil. The major advantages of maize dry milling are the lower capital costs as compared to wet milling [3].

2.2 Wet milling

In developed countries like the USA, the major utilisation has been wet milling. The two most important products of wet milling are high-fructose corn syrup and ethanol.

2.3 Alkali processing

In this process, maize is cooked with water and lime at 90°C for 50 minutes then steeped for 14 hours before being washed with fresh water to remove residual alkali and other waste materials from maize.

3. Products of maize

There are varieties of products that can be derived after application of food processing methods like dry milling, wet milling and alkali processing to make it consumable for human beings as discussed above. Commonly, maize is used to make flour, oil, starch, grits, flakes, popcorn, etc. [4]. Few very popular products derived from maize are discussed below.

3.1 Degerminated flour

This consists mostly of the endosperm and has content of B vitamins. It is used by brewers as a starch medium for the action of barley malt in the preparation of wort for the production of beer. It is also used to make chapatti or bread in the northern region of India. The flour is supplemented with green leafy vegetables to make it more nutritious and healthy. The chapatti is famous around the Punjab as '*makki di roti*' that is served with well-cooked mustard leaves along with butter.

3.2 Corn germ oil

It can be obtained by solvent extraction. Maize oil has become a highly desired vegetable oil owing to its relatively high level of linolenic fatty acid and its excellent flavour. The fat content of maize is 3.6%, and oil extracted from it can be refined to produce a high-quality vegetable oil for cooking or food use.

3.3 Popcorn

A particular variety of corn is used to make popcorn and most famous food items derived from maize. To make popcorn a hard corneous endosperm is desirable. Other desirable traits of popcorn are good flavour, tenderness, the absence of objectionable hulls and high popping expansion. Moreover, the moisture content recommended is 13.5% for the best popping expansion. The popping of corn is a method of starch cookery. As the kernels of popcorn are heated, the water vapour within them expands, increasing the pressure until it is sufficient to make the kernels explode or 'pop' [5]. Popping can be done with or without fat as well. Ready-to-cook popcorns are also easily available that are enriched with both nutrition and taste. Thus it can be added as supplementary snacks in the diet of malnutrition children.

3.4 Corn starch

It is the most widely used product obtained from maize. It is made by a process of wet milling in which the hull and germ are removed and the corn ground and mixed with water. The semi-liquid material is separated by passing it over sieves or centrifuging it. The starch settles out while most of the proteins are suspended. The starch is then washed, dried and powdered. Corn starch is widely used because it is inexpensive, lacks characteristic flavour and cooks to a smooth and almost clear paste in water or other clear liquids and is superior to wheat flour or potato starch. Corn starch flavoured with vanilla and containing edible colours is solid as custard powder.

3.5 Cornflakes

The whole grain is crushed between the large metal rollers to remove the bran from the outer layer and then mixed with seasoning agents (salt, sugar, flavours and fortified minerals) and water in a large rotating pressure cooker. The physiochemical properties like time, temperature and speed of rotation vary with the type of grain being cooked. The cooked grain is moved to a conveyor belt which passes through a drying oven. In this process, soft and solid mass is obtained which can be moulded into desired shapes. Then these cooked grains are allowed to cool, and stabilising the moisture content is known as 'tempering'. Then the tempered grains are flattened between large metal rollers under tons of pressure, and the resulting flakes are further conveyed to ovens with blast of very hot air to remove reaming moisture and to toast them to desirable flavours. Cornflakes are also processed from extruded pellets in a similar way.

4. Nutritive value of maize

4.1 Macronutrients

Maize provides approximately 1400 Kcal/100 g (on a dry basis) of energy that is sufficient to maintain the equilibrium. This energy is also used to perform different types of physiological task. Maize or corn can be consumed as a source of energy in the form of breakfast cereals as cornflakes, chapattis, tortillas, etc. Maize also contains an appreciable amount of fat content that helps in the carrier of fat-soluble vitamins A, D, E and K. The presence of fat in maize or corn is responsible for much of the texture and flavour of food. Thus it helps in increasing the palatability.

The fat content beneath the skin known as the subcutaneous fat also serves as an insulating material for the body and is effective in preventing heat loss. Moreover, fat content also acts as a body reservoir for energy conservation purpose.

Another important component in maize after fat is dietary fibre and is defined as the portion of food derived from plant cell, which is resistant to hydrolysis or digestion by the elementary enzyme system in human beings. However, some of the bacteria in the large intestine can degrade some components of fibre releasing products that can be absorbed into the body and also used as a source of energy. Crude fibre is the residue remaining after the treatment with hot sulphuric acid, alkali and alcohol. The major component of crude fibre is a polysaccharide called cellulose and a part of dietary fibre. Insoluble fibres are indigestible and insoluble in water, while soluble fibres are indigestible but soluble in water. Total fibre is the sum of insoluble and soluble fibres. Dietary fibre is isolated and extracted from a synthetic fibre that has proven health benefits. Resistant starch also functions as dietary fibre [6–8].

$$\text{Total fibre} = \text{dietary fibre} + \text{functional fibre}$$

The effect of fibre on the gastrointestinal tract (**Table 1**) is influenced by the characteristics of the fibre itself, the particle size, the interaction between fibre and other dietary components and the bacteria flora. Maize also contains a significant quantity of insoluble fibre found in the cell wall of the constituent [9].

The insoluble fibre present in maize or corn has a physiological effect in preventing constipation, diverticulitis and even cancer of the large intestine as presented in **Table 2**.

Maize is also considered as a booster of nutrient like carbohydrates, fats, proteins and insoluble fibres that helps in providing sufficient energy to meet the human daily dietary requirements [10]. The proximate composition of maize is presented in **Table 3**.

4.1.1 Protein

Maize contains 8–11% of protein that is made from different components like albumin, globulin, nonnitrogen substance, prolamin, etc. The quality of maize protein depends upon its agronomic practices and genotype as well. The quality of maize protein is not of good quality as compared to other cereal grains like rice, wheat, barley, etc. Recent researches have shown that with genetic modification, the quality of maize protein can be improved. The maize protein is known as zein that is lack of essential amino acids tryptophan and lysine. The opaque-2 gene is also helpful in reducing the concentration of zein up to 30% and improves the quality protein maize (QPM). The protein content present in maize helps in the growth

Site	Activity
Mouth	Stimulate saliva secretion
Stomach	Dilutes contents, delays gastric emptying
Small intestine	Dilutes content, delay absorption
Large intestine	Dilutes contents, forms substrate for bacteria, traps water, binds cation, soften stools, prevents straining

Source: Raninen et al. [9].

Table 1.
Influence of dietary fibre on the gastrointestinal tract.

Disease	Physiological mechanism
Constipation Diverticulitis Irritable bowel syndrome Varicose veins Haemorrhoids	• Increase the water-holding capacity • Increases the stool weight • Reduces the transit time • Enhances gastric motility • Volatile fatty acids which are released by bacteria having a laxative effect • Faster bowel emptying due to increased intraluminal mass bulk • Decreases intracolonic pressure
Cancer of the large intestine	• Changes in the population of microbes in the GI tract • Increases binding of intestinal bile acids • Food residues remain in the colon for less time for carcinogen to be absorbed • Increases stool weight and volume • Increases frequency of defaecation • Bulk and water of the faeces may dilute the carcinogen to a nontoxic level • Fibre-induced effects on faecal enzymes • Production and distribution of short-chain fatty acids in the colon resulting in pH modifications • Increases on bile acids and mutagens in the colon • Adsorbing cancer-producing hydrocarbons

Source: Raninen et al. [9].

Table 2.
Role of dietary fibre in preventing and managing diseases.

S. no.	List of nutrients	Nutritive value*
1	Moisture	9.26 ± 0.55
2	Protein (g)	8.80 ± 0.49
3	Ash (g)	1.17 ± 0.16
4	Fat (g)	3.77 ± 0.48
5	Total fibre (g)	12.24 ± 0.93
6	Insoluble fibre (g)	11.29 ± 0.85
7	Soluble fibre (g)	0.94 ± 0.18
8	Carbohydrates (g)	64.77 ± 1.58
9	Energy (KJ)	1398 ± 25

Source: Longvah et al. [6], Indian Food Composition Tables, Government of India.
**All the values are presented as per 100 grammes of edible portion.*

Table 3.
*Nutritive value of proximate content of maize, dry (*Zea mays*).*

and maintenance of tissues, formation of essential body compounds, transport of nutrients, regulation of water balance, maintenance of appropriate pH, defence and detoxification as well.

4.1.2 Essential amino acids

These amino acids cannot be synthesised by the body at a sufficient rate to meet the body requirement for optimum growth and development. The human body has certain limited powers of converting one amino acid into another. This is

Essential amino acids		Conditionally essential amino acids		Non-essential amino acids	
Amino acids	**Nutritive value***	**Amino acids**	**Nutritive value**	**Amino acids**	**Nutritive value***
Histidine	2.70 ± 0.21	Arginine	4.20 ± 0.24	Alanine	7.73 ± 0.46
Isoleucine	3.67 ± 0.22	Cysteine	1.55 ± 0.14	Asparagine	—
Leucine	12.24 ± 0.57	Glycine	3.27 ± 0.15	Aspartic acid	6.55 ± 0.59
Lysine	2.64 ± 0.18	Proline	7.88 ± 0.71	Glutamic acid	19.39 ± 0.70
Methionine	2.10 ± 0.17	Tyrosine	3.71 ± 0.18	Glutamine	—
Phenylalanine	5.14 ± 0.29			Serine	4.58 ± 0.44
Threonine	3.23 ± 0.29			Selenocysteine	—
Tryptophan	0.57 ± 0.12			Pyrrolysine	—
Valine	5.41 ± 0.71				

Source: Longvah et al. [6], Indian Food Composition Tables, Government of India.
**All the values are presented as per 100 grammes of edible portion.*

Table 4.
*Essential and non-essential amino acid profile (g) of maize, dry (*Zea mays*).*

achieved in the liver by the process of transamination, whereby an amino group is shifted from one molecule to another under the influence of amino transferases, the coenzyme of which is pyridoxal phosphate. The inability to synthesize the carbon skeleton of these amino acids is the probable reason why they are dietary essentials. There are nine essential amino acids that are required for a human body to perform various functions (**Table 4**).

4.1.3 Conditionally essential amino acids

These are needed in the diet unless abundant amounts of their precursors are available for their synthesis. The newborn may not have enzymes in adequate amounts to synthesise non-essential amino acid, or in intestinal metabolic dysfunction, arginine may not be synthesized. Hence it becomes a conditionally essential amino acid. Amino nitrogen is not freely interchanged between all amino acids. The precursors of conditionally essential amino acids are mentioned in **Table 4**.

4.1.4 Non-essential amino acids

Non-essential amino acids are the ones that the body can make in adequate amount if nitrogen is available in the diet. They are non-essential only in the sense that they are not essential components of the diet as discussed in **Table 4**.

4.1.5 Starch

The main portion of maize grin is starch that provides more than 70% weight to its cereal kernel. Starch in maize is composed of two glucose polymers mainly amylose that contributes to 30% of its starch content and the rest of the content is made from amylose pectin (70%). Waxy maize is composed of 100% amylopectin content. Due to the pectin content, maize has a branch-type structure. The monosaccharide present in maize is comprised of glucose and fructose, and the disaccharide is comprised of sucrose in a little amount. The starch and sugar content of maize is presented in **Table 5**.

S. no.	List of nutrients	Nutritive value*
1	Total available CHO	61.01 ± 0.76
2	Total starch	59.35 ± 0.83
3	Fructose	0.16 ± 0.03
4	Glucose	0.80 ± 0.01
6	Sucrose	0.70 ± 0.03
6	Total free sugars	1.66 ± 0.04

Source: Longvah et al. [6], Indian Food Composition Tables, Government of India.
**All the values are presented as per 100 grammes of edible portion.*

Table 5.
Starch and sugar content (g) of maize, dry (Zea mays)*.*

S. no.	List of nutrients	Nutritive value*
1	Thiamine (B1) (mg)	0.35 ± 0.039
2	Riboflavin (B2) (mg)	0.14 ± 0.014
3	Niacin (B3) (mg)	2.10 ± 0.09
4	Pantothenic acid (B5) (mg)	0.27 ± 0.02
5	Total B6 (mg)	0.28 ± 0.023
6	Biotin (B7) (mg)	0.70 ± 0.06
7	Total folates (B9) (mg)	39.42 ± 3.13

Source: Longvah et al. [6], Indian Food Composition Tables, Government of India.
**All the values are presented as per 100 grammes of edible portion.*

Table 6.
B-complex nutritive content of maize, dry (Zea mays)*.*

4.2 B-complex vitamins

Maize is also enriched with B-complex vitamins that play a vital role in growth, healthy skin, heart, hair, brain, digestion, nails and dementia as well. Maize products can be used in the daily diet of coeliac patients to improve their health status [11]. People with coeliac disease cannot absorb gluten due to an abnormal immune reaction that occurs in the small intestine. So the only cure is consumption of gluten-free diet that helps in improving the gastrointestinal function [12]. Maize is enriched with thiamine, riboflavin, niacin, pantothenic acid, pyridoxine and folic acid as well. The nutritional content of B-complex vitamin present in maize is discussed in **Table 6**.

The B-complex vitamin present in maize is of water-soluble nature and found in the aleurone layer of the kernel. The processing method has significant direct relationship with the amount of vitamin present in maize. Moreover, niacin deficiency causes pellagra that is also directly related with maize.

4.3 Fat-soluble vitamins

Maize contains fat-soluble vitamins that is comprised of provitamin A, carotenoids, lutein, zeaxanthin, ergocalcifeol, tocopherol, phylloquinones (**Table 7**), etc. that have a unique role in preventing both ageing and cancer. These fat-soluble vitamins (A, D, E, K) act as antioxidants and scavenge the free radicals that help in protection against different types of cancer. The content of fat-soluble vitamin

S. no.	List of nutrients	Nutritive value*
1	Lutein (μg)	186 ± 19.4
2	Zeaxanthin (μg)	42.4 ± 15.7
3	B-Cryptoxanthin (μg)	110 ± 10.1
4	β-Carotene (μg)	186 ± 19.2
5	Total carotenoids (μg)	893 ± 154
6	Ergocalciferol (μg)	33.60 ± 2.82
7	Tocopherol-alpha (mg)	0.21 ± 0.04
8	Tocopherol-gamma (mg)	1.29 ± 0.17
9	Tocopherol-delta (mg)	0.38 ± 0.05
10	Tocotrienol-alpha (mg)	0.05 ± 0.00
11	α-Tocopherol, vitamin E (mg)	0.36 ± 0.03
12	Phylloquinones (vitamin K (μg)	2.50 ± 0.76

Source: Longvah et al. [6]s, Indian Food Composition Tables, Government of India.
**All the values are presented as per 100 grammes of edible portion.*

Table 7.
*Nutritive content of fat-soluble vitamin in maize, dry (*Zea mays*).*

S. no.	List of fatty acids	Nutritive value*
1	Palmitic (C16:0)	363 ± 4.6
2	Stearic (C18:0)	42.45 ± 2.76
3	Arachidic (C20:0)	7.14 ± 0.95
4	Oleic (C18:1)	700 ± 17.9
5	Eicosaenoic (C20:1)	6.62 ± 0.74
6	Linoleic (C18:2)	1565 ± 18.2
7	α-Linolenic (C18:3)	40.76 ± 2.43
8	Total saturated fatty acids (TSFA)	413 ± 5.6
9	Total monounsaturated fatty acids (TMUFA)	706 ± 17.4
10	Total polyunsaturated fatty acids (TPUFA)	1606 ± 18.5

Source: Longvah et al. [6], Indian Food Composition Tables, Government of India.
**All the values are presented as per 100 grammes of edible portion.*

Table 8.
*Essential fatty acid profile (mg) of maize, dry (*Zea mays*).*

depends upon the genotype of maize whether it is fortified or not, that is, yellow maize is enriched with different types of carotenoid pigment due to its genotype, while white maize is deficient in carotenoid content due to absence of this genotype.

Majority of the carotenoid contents are present in the hard endosperm of maize kernel and the rest in small quantity in the germ. The ergocalciferol content present in maize helps in the bone formation and tocopherol (α, β, γ) in anti-ageing and cosmetic products. As an antioxidant, tocopherol (vitamin E) helps in protecting different types of skin cancer. Phylloquinone (vitamin K) helps in the blood clotting when an accident or injury happens. It has anticoagulating properties. The following tables discussed the nutritive value of fat-soluble vitamin content present in maize (**Tables 7–12**).

S. no.	List of nutrients	Nutritive value*
1	Aluminium (Al)	2.82 ± 0.16
2	Calcium (Ca)	8.91 ± 0.61
3	Chromium (Cr)	0.010 ± 0.006
4	Cobalt (Co)	0.010 ± 0.003
5	Copper (Cu)	0.45 ± 0.23
6	Iron (Fe)	2.49 ± 0.32
7	Lithium (Li)	0.002 ± 0.001

Source: Longvah et al. [6], Indian Food Composition Tables, Government of India.
**All the values are presented as per 100 grammes of edible portion.*

Table 9.
*Minerals and trace content (mg) of maize, dry (*Zea mays*).*

S. no.	List of nutrients	Nutritive value*
1	Total oxalate	15.26 ± 1.78
2	Insoluble oxalate	14.19 ± 1.30
3	Soluble oxalate	2.73 ± 1.34
4	Fumaric acid	0.66 ± 0.20
5	Malic acid	0.93 ± 0.50
6	Quinic acid	0.84 ± 0.07
7	Succinic acid	1.50 ± 0.23
8	Tartaric acid	0.94 ± 0.05

Source: Longvah et al. [6], Indian Food Composition Tables, Government of India.
**All the values are presented as per 100 grammes of edible portion.*

Table 10.
*Total organic acid content (mg) of maize, dry (*Zea mays*).*

4.3.1 Essential fatty acids (EFA)

The oil content of maize is extracted from the germ part which is genetically modified with an average range of 3–18%. Three classes of fatty acids are described according to the number of double bonds between the carbon atoms as described in **Table 8**. In saturated fatty acids, there are none; in an unsaturated fatty acid, there may be one (monoenoic or monounsaturated fatty acids) or two or more (polyenoic or polyunsaturated fatty acids) double bonds. Corn oil is enriched with PUFA (polyunsaturated fatty acid) and MUFA (monounsaturated fatty acid) while having low content of SFA (saturated fatty acid). SFA comprises of palmitic, stearic and arachidic acids. PUFA contains linoleic, α-linolenic, arachidonic and eicosaenoic acids that help in maintaining healthy skin and vision, strong immune system and optimum growth and development. Moreover, it has also anti-inflammatory properties and reduces the production of interleukin-1 and tumour necrosis factor (TNF) by downregulating inflammatory response. It is also responsible for the formation of prostaglandins that are found in every single cell of the body and helps in regulating cell activities including transmission of genetic information from generation to generation.

This is rare in human beings that deficiency of fatty acid occurs. It has been reported, however, in patients fed solely by vein (total parenteral nutrition (TPN))

S. no.	List of nutrients	Nutritive value*
	Phenols	
1	3,4-Dihydroxybenzoic acid	0.07 ± 0.02
2	Protocatechuic acid	2.93 ± 0.42
3	Vanillic acid	2.96 ± 0.44
4	p-Coumaric acid	2.84 ± 0.36
5	Caffeic acid	2.91 ± 0.32
6	Chlorogenic acid	1.01 ± 0.45
7	Ferulic acid	1.43 ± 0.09
8	Total polyphenols	32.92 ± 3.85
	Phytosterol	
1	Campesterol	12.49 ± 0.24
2	Stigmasterol	4.22 ± 0.18
3	β-Sitosterol	87.70 ± 2.61
	Phytate	646 ± 19.4

Source: Longvah et al. [6], Indian Food Composition Tables, Government of India.
**All the values are presented as per 100 grammes of edible portion.*

Table 11.
*Total polyphenol, phytosterol and phytate contents (mg) of maize, dry (*Zea mays*).*

S. no.	Nutrients	Deficiency
1	Energy, protein	Underweight, marasmus, kwashiorkor
2	Fibre	Constipation, diverticulitis
3	Calcium	Rickets, osteomalacia
4	Iron	Anaemia
5	Vitamin A	Night blindness
6	Thiamine	Pain in the calf muscle, weakness of the heart muscle
7	Niacin	Dementia, diarrhoea, dermatitis
8	Pyridoxine	Angular stomatitis
9	Folic acid	Megaloblastic anaemia
10	Antioxidants	Decreased immunity

Source: Srilakshmi, [17], Text Book on Nutrition Science.

Table 12.
Nutrient deficiency symptoms related to maize.

for long times without fat emulsions. EFA deficiency can occur in fat malabsorption and occasionally in protein-calorie malnutrition, where there is a deficiency of fat calories.

4.4 Mineral and trace element

The majority of minerals and trace elements of maize is present in germ portion and very few in endosperms. Phosphorus is found in the embryo portion of the maize. Environmental factors strongly influence the quality and quantity of mineral content present in maize. Maize is having a low content of mineral and trace element (**Table 9**) as compared with other cereal grains. The minerals present in maize have a vital role in bone development, tooth formation, haemoglobin formation, growth regulation, regulation of acid–base balance of the body, facilitation of energy transactions, absorption and transport of nutrients and metabolism of carbohydrates, proteins and fats. These minerals also act as a cofactor and regulator of biochemical reactions for blood clotting, contraction of muscles, releases of insulin and parathyroid and calcitonin hormones as well. Furthermore, these minerals play vital role in the growth, development and formation of red blood cells in human system. The chromium content of an adult body is estimated to be 6 mg and potentiates insulin action. The mineral and trace element content of maize is discussed below.

4.5 Organic acid content

There are a number of organic acids present in nature like formic, malic, succinic, etc. Organic acids help in building the carboxylic acids which can alter the physiology of bacteria and cause metabolic disorders that prevent their proliferation and death. Organic acids are not fixed in one state, and supplementation of its higher doses in animal feed helps them to gain body weight and improves feed conversion ratio by reducing the colonisation of pathogens in the intestine. The total organic content of maize is discussed in **Table 10**.

4.6 Antioxidants

Food polyphenols (**Table 11**) are ubiquitous components and have an antioxidant mechanism involved in fighting free radical damage by interaction of ascorbic acid and glutathione (GSH) with oxidants and oxidising agents. Scavenging of free radicals and single oxygen through food polyphenols (Vitamin E, ascorbic acid, ß-carotene and superoxide dismutase) by reduction of hydroperoxides, glutathione peroxidases (GSHPx) and catalase enzymes as well. Food polyphenols also act as chelating agent by binding with transition metals that cause cellular damage [13]. Thermal processing deteriorates the quality of maize grains due to leaching of water-soluble polyphenols into brine or sugar solution. The effects of processing method cause alteration in the structure, chemical composition and nutritional value of the food products like canned sweet corn, tortillas, chips, etc. [14].

Recently, the industry has focused attention to plant matrices rich in phytosterols and phytostanols for their ability to reduce serum cholesterol levels. Therefore, the objective of this study was to examine the phytosterol and phytostanol contents of different fractions (endosperm, pericarp, germ) of corn kernel. The phytosterols are found in the endosperm, pericarp and germ portion of corn kernel. The germ portion contains 25–31% of oil as compared to other fractions. Corn oil is enriched with ß-sitosterol (62–69%), followed by campesterol (11–18%) and stigmasterol (5–13%). Processing of maize, especially during roasting, results in the loss of phytate content and increase of the availability of minerals. For example, baking chapattis from maize helps in the reduction of phytates and improves the nutritional quality of maize. Due to the emerging field of nutraceuticals, the phytochemicals derived from maize have achieved great attention. The antioxidant capacity in terms of DPPH radical scavenging activity of maize (139 mg/100 g) is quite high as compared to other cereal and pulse grains except finger millet (173 mg/100 g). This antioxidant activity of maize helps in protecting different types of degenerative diseases [15].

4.7 Relationship of maize with health

Being comparatively inexpensive, maize as a stable diet contributes to most of the caloric requirement. It is also an excellent source of starch and B-complex vitamins along with antioxidants such as different types of polyphenols [16]. It also contributes to satiety and is used as a main dish in the diet. No meal can be prepared from cereals. Maize is also used as a thickening agent as a corn flour in custards and puddings. Maize as a thickening agent used in the preparation of different types of sausages as well.

Maize is also consumed as a ready-to-eat food in the form of cornflakes with milk as a healthy breakfast [17]. The fibre present in the maize helps to lower cholesterol levels and reduce the risk of colon cancer (**Table 12**). Moreover, it is also useful for anaemic, haemorrhoid, cardiac and diabetic patients due to significant nutritional value of macro- and micronutrients in it. It is also helpful in the metabolism of carbohydrates due to the presence of thiamine in it [18].

Author details

Shikha Bathla[1*], Manpreet Jaidka[1] and Ramanjit Kaur[2]

1 Punjab Agricultural University, Ludhiana, Punjab, India

2 ICAR-IARI, New Delhi, India

*Address all correspondence to: shikha_bathla@yahoo.com

References

[1] Shah TR, Prasad K, Kumar P. Maize-a potential source of human nutrition and health: A review. Cogent Food & Agriculture. 2016;**2**:1-9

[2] Kumar D, Jhariya NA. Nutritional, medicinal and economical importance of corn: A mini review. Research Journal of Pharmaceutical Sciences. 2013;**2**:07-08

[3] Singh N, Richa B, Rhythm G, Mukti G, Jaghmohan S. Physico-chemical, thermal and pasting properties of fractions obtained during three successive reduction milling of different corn types. Food Chemistry. 2009;**113**:71-77. DOI: 10.1016/j.foodchem.2008.07.023

[4] Locatelli S, Berardo N. Chemical composition and phytosterols profile of degermed maize products derived from wet and dry milling. Maydica. 2014;**59**:261-266

[5] Suri DJ, Sherry A. Effects of different processing methods on the micronutrient and phytochemical contents of maize: From A to Z. Comprehensive Reviews in Food Science and Food Safety. 2016;**15**:1541-4337

[6] Longvah T, Ananthan R, Bhaskarachary K, Venkaiah K. Indian Food Composition Tables. Hyderabad, Telangana State, India: National Institute of Nutrition, Indian Council of Medical Research, Department of Health Research, Ministry of Health & Family Welfare, Government of India; 2017

[7] Willis HJ, Eldridge AL, Beiseigel J, Thomas W, Slavin J. Greater satiety response with resistant starch and corn bran in human subjects. Nutrition Research. 2009;**29**:100-105. DOI: 10.1016/j.nutres.2009.01.004

[8] Higgins JA. Resistant starch: Metabolic effects and potential health benefits. Journal of AOAC International. 2004;**87**:761-768

[9] Raninen K, Lappi J, Mykkanen H, Poutanen K. Dietary fibre type reflects physiological functionality: Comparison of grain fibre, inulin and polydextrose. Nutrition Reviews. 2011;**69**(1):9-21. DOI: 10.1111/j.1753-4887.2010.00358.x

[10] David-Jenkins JA. High-complex carbohydrates or Lente carbohydrates foods. American Journal of Medicine. 2002;**113**(9):30S-37S. DOI: 10.1016/s0002-9343(01)00989-5

[11] Kataria R. Proximate nutritional evaluation of maize and Rice-gluten free cereal. Journal of Nursing and Health Science. 2014;**3**(2):01-06

[12] Gallagher E, Gormley TR, Arendt EK. Recent advances in the formulation of gluten free cereal-based products and trends. Food Science and Technology. 2004;**15**:143-152

[13] Shahidi F. Nutraceuticals and functional foods: Whole versus processed foods. Trends in Food Science and Technology. 2009;**20**:376-387

[14] Liu RH. Whole grain phytochemicals and health. Journal of Cereal Science. 2007;**46**:207-219. DOI: 10.1016/j.jcs.2007.06.010

[15] Liu RH. Potential synergy of phytochemicals in cancer prevention: Mechanism of action. Journal of Nutrition. 2004;**134**:3479-3485

[16] Lopez-Martinez LX, Oliart-Ros RM, Valerio-Alfaro G, Lee CH, Parkin KL, Garcia HS. Antioxidant activity, phenolic compounds and anthocyanins content of eighteen strains of Mexican maize. LWT - Food Science and Technology. 2009;**42**:1187-1192

[17] Srilakshmi B. Nutrition Science. 5th ed. New Delhi: Published globally by New Age International (P) Limited, Publishers; 2016

[18] Srilakshmi B. Food Science. 6th ed. New Delhi: Published globally by New Age International (P) Limited, Publishers; 2015

Chapter 3

Quality Protein Maize: An Alternative Food to Mitigate Protein Deficiency in Developing Countries

S.R. Krishna Motukuri

Abstract

Maize (*Zea mays* L.) plays a significant role in human nutrition and animal feed. After the discovery of opaque-2 mutants in maize, that produces with enhanced levels of lysine and tryptophan. Quality protein maize (QPM) holds superior nutritional value and is essentially exchangeable with normal maize. The increasing use of maize as feed, increasing interest of the consumers in nutritionally enriched products and rising demand for maize seed are the core driving forces behind emerging importance of maize crop in India. Protein malnutrition is a serious global issue demanding huge resources on healthcare. The problem can be addressed to a considerable extent by shifting to quality protein maize diet. The development of QPM hybrids through advanced breeding approach like molecular marker-assisted breeding was adopted. It could solve the issue related to protein deficiency in developing countries.

Keywords: maize, opaque-2, QPM, protein deficiency, marker-assisted breeding

1. Introduction

Improvement of protein quality of maize incorporated the mutant gene called opaque-2, thus leading to the development of quality protein maize (QPM). Several natural mutants, which confer the highest lysine and tryptophan levels, had been identified in the 1960s and 1970s, i.e. opaque-2, opaque-6, opaque-7, floury-2 and floury-3 [1]. QPMs are having more quantity of lysine and tryptophan and lesser quantity of leucine and isoleucine. Baby corn, sweet corn, popcorn, waxy corn and high oil corn were targeted to develop quality protein maize [2]. QPM hybrids with different kernel colors have been developed and are released in India for their cultivation in various agroclimatic conditions. The technology involved in the production of QPM and normal maize are the same, but QPMs should be grown separately to maintain its purity.

In the QPM, recessive opaque-2 (o2) allele has been successfully utilized in the conversion breeding program for increasing the quality of protein in maize [3]. Primarily, maize varieties with o2 mutation were not chosen by farmers and consumers, because of opaque endosperm. Opaque-2 mutant is susceptible to pests and diseases, and it also undergoes grain breakage during milling [4]. Endosperm modifier genes, which present hard endosperm in the o2 background, were developed at the International Maize and Wheat Improvement Center (CIMMYT), Mexico [5], and University of Natal, South Africa [6]. This leads to the development of nutritionally enriched hard endosperm maize, widely known as 'quality protein maize' [3].

2. Nutrition deficiency and related challenges

With the increasing world population, enhancing the production of food and nutritional quality of staple crops is the strategy to address the emerging food crises [7]. A food crisis causes multidimensional effects on human nutrition, and it causes malnutrition. It also has effects on the supply of food quantity and quality of food. In the last two decades, these problems have been tried to be solved to reduce the proportion of the world's malnourished population [8]. Protein deficiency malnutrition has emerged as a major nutritional problem, particularly in the developing countries [9]. In the developing countries, cereals play an important source of dietary protein for humans, which comprise 70% of the protein intake [10]. Maize is the world's third primary cereal crop, which is an important protein source used as food and feed for humans and animals and also used in corn starch industry, corn oil production, etc. [11]. QPM has more quantity of carbohydrates, fats, proteins, vitamins and minerals. It is also called as a 'poor man's cereal crop'. In developing countries like Africa and Latin America, as the animal protein is very limited and expensive, which results in being unavailable to a vast sector of the population, maize grains provide about 15–56% of total daily calories in people's diets [12]. Nearly 9.09 million hectares were allocated to cultivate maize, which produces nearly 24.26 million tons in India and can be cultivated throughout the year [13]. Maize proteins consists just 1.81 and 0.35% of lysine and tryptophan content, respectively, which is very low compared with the Food and Agriculture Organization (FAO) recommendation. From the human nutrition perspective, lysine and tryptophan are the most considerable limiting amino acid in the maize endosperm protein. Thus humans and other monogastric animals should include other alternative sources of lysine and tryptophan in their healthy diets [14]. Babies fed on normal maize without any protein supplements suffer from malnutrition and develop Kwashiorkor disease [15]. In this context, the International Maize and Wheat Improvement Center (CIMMYT) and the International Institute of Tropical Agriculture (IITA) are developing varieties to improve the protein quality of maize by incorporating the opaque-2, along with modifier genes, thus increasing the amount of lysine (>4.0%) and tryptophan (>0.8%) contents in the whole grain compared with normal maize [16]. Maize cultivars containing high yield with increasing levels of lysine and tryptophan and having the kernel structure of conventional maize have the potential to reduce the malnutrition [14].

3. Storage proteins in QPM

The mature maize kernel consists of a germ, pericarp and endosperm. An endosperm consists of 90% starch which is a source of concentrated energy and 10% protein which include albumins, globulins, zein and glutelin out of which zein consists 50–70% of total proportion [10]. Zeins are the important storage proteins; these forms as deposit on rough endoplasmic reticulum-delimited protein bodies (PBs) [17]. During the maturation of kernel, these protein bodies become densely packed between starch grains in the vitreous regions of the endosperm [18]. Zeins are a group of four structurally distinct alcohol-soluble proteins (α-zein, β-zein, γ-zein and δ-zein) [17] present only in seeds' endosperm and playing a key role in storing and supplying N, C and S to the germinating seedling [12]. Among those zein proteins, α-zeins and δ-zeins are deposited in the central region, and γ-zeins and β-zeins were deposited in the outer region of protein bodies [19]. The zein fractions are rich in cysteine and methionine amino acids, and it also consists of glutamine, leucine and proline and is completely devoid of two important essential amino acids lysine and tryptophan, whereas other proteins consist of these amino acids in large quantities [20]. The zein synthesis serves as a model system to study coordinated genetic regulation of several genes expressed at very high levels at a specific developmental stage. Suppression of zein fraction without drastically altering the contribution of other fractions could be, thus, seen as a feasible approach to bring about improvements in the amino acid balance in maize grain [12].

3.1 Zein gene

Zein is a class of prolamin proteins that are mainly present in maize. All the zein polypeptides are products of different structural genes [21]. Most of the prolamin genes have a promoter element called the endosperm or prolamin box. The promoter element is present about the 300 base pairs upstream of the translation start codon and has a conserved 15-bp element that contains the 7-bp endosperm motif (TGTAAAG) [22]. This endosperm motif acts as a tissue-specific enhancer in Mr. 22,000 gene promoters [23].

Genetic analysis of o2 modifiers revealed several quantitative trait loci (QTLs) dispersed on the chromosomes. These identified QTLs were correlated with the 27-kDa γ-zein gene expression and protein quantity in QPM [24]. The 27-kDa γ-zein gene expression is not under the control of the o2 protein [25]. The o2 modifier genes involved in the 27-kDa γ-zein gene expressions are observed in two different QTLs. The first of these is associated with increased expression [26]. Single copy of γ-zein genes encodes the 50, 27 and 16-kDa proteins, which were observed in the B73 genome [27]. Based on the allotetraploidization and protein-sequence similarity, both 27 and 16 kDa γ-zein genes originated from a common progenitor [28]. It is about 20–25% of total zeins; the low abundance 50-kDa γ-zein gene has low similarity with other two γ-zein genes [27]. The γRNAi and βRNAi were involved in maize kernel opacity to increase the intensification. It reveals that opacity was not involved in reducing the thickness of the opaque-2-mutated endosperm; it is due to partial arrangement of starch granules in the endosperm [29]. Although discrete protein bodies were observed in endosperm cells, honeycomb-like masses of protein bodies were observed. It indicates that different zeins have played an important role in the endosperm development.

4. Nutrition analysis of QPMs

Generally, quality of protein nutrition was estimated by composition of amino acids, digestibility and amino acid requirement to consume the protein. The QPMs are reported to have increased levels of lysine and tryptophan in the endosperm protein, which enhances the biological value of protein similar to the milk protein. It has brought about great hope in the effort to improve human nutrition [30]. Firstly there is a significant difference in the QPM kernel when compared to normal maize kernel. Kernel hardness was determined by calculating floatation index where it is 57% for QPM, whereas for normal maize, it is 19.7%. The whole kernel protein was 13.15% in QPMs with contribution of 8.6 and 13.88% from endosperm and germ, whereas it is 9.25% in normal maize with contribution of 7.9 and 1.28% from endosperm and germ, respectively [18]. An improvement of protein quality has been correlated with the presence of the opaque-2 mutant gene [31]. Crude protein of QPM was higher than the normal maize, and the proportional contribution of the germ is lower in QPMs than with normal varieties. These structural and biochemical changes that happen in the kernel lead to the modifications of the protein profile, both in content and structure, and therefore on the functionality of the protein extracted from QPM [30]. Based on the chemical component analysis, QPM whole kernels showed highest protein content compared with normal maize [32].

5. Efforts in enhancing QPM production

5.1 Genetics of QPM

QPM contains the mutation at opaque-2 loci, which changes the protein composition of the maize endosperm, resulting in increased concentrations of lysine and tryptophan [33]. The increase in concentration (60–100%) of these two essential amino acids increased the biological value of QPM (80%), when compared to normal maize (40–57%) [34]. The biological value of cow milk protein was about 90%, whereas QPM has about 80% value [35].

QTL mapping of o2 modifiers insights that it encodes that the 27-kDa - zein protein and it is observed on chromosome-7 long arm [36]. The function of the 27-kDa zein protein in the formation of vitreous endosperm was revealed when the protein quantity increased threefold in QPM compared with soft opaque-2 mutant [37]. An increase in the number of zein proteins and their compaction between starch grains is partially involved in endosperm modification in QPM [38]. The o2 modifier genes have complexity in inheritance [12]; it reveals that several other loci control the formation of a vitreous kernel in QPM. For identifying the other factors linked to the endosperm modification, [39] performed a proteomic study of the non-zein proteins, and it was observed that the quantity of a starch synthesis enzyme and the amylopectin branching structure are changed in QPM. It is supported that QPM starch expands more than normal maize. It reveals that suppression of the opaque endosperm in QPM was associated with the starch grain properties.

Maize protein quantity can be enhanced with the opaque-2 (o2) mutation, which increases the lysine and tryptophan levels by decreasing the synthesis of zeins. The QPM utilization mainly restricts due to chalky and soft texture kernels [3]. The quality protein maize was developed based on introgression of opaque-2 QTLs, called o2 modifiers which convert to hard and vitreous endosperm [40]. QPM development has significantly improved the status of nutrient-deficient people who suffer from malnutrition and protein energy deficiency in the developing countries [41].

6. Breading efforts in QPM

Although QPM breeding has been practiced for more than 60 years, genetic mechanism and genetic components controlling endosperm modification are not clearly understood. Opaque-2 (o_2) modifier loci have been distributed on six chromosomes [26]. The opaque-2 modification is positively correlated with 27-kDa γ-zein in an F_2 population and recombinant inbred lines (RILs), which are produced through crosses between QPM and an o2 mutant as parents [42]. Gene silencing or deletion of γ-zeins eliminates 27-kDa γ-zein expression, and it eliminates the formation of vitreous endosperm [43]. Zein proteins are stored at rough endoplasmic reticulum-retained protein bodies in the endosperm [44]. For protein body formation, 27-kDa γ-zein, 16-kDa γ-zein and 15-kDa β-zein plays an important role in initiation and stabilization [19]. Zein gene knockout studies in QPM showed irregular, clumped protein bodies in lesser number and an opaque phenotype [29].

Worldwide different agricultural research centers are showing significant progress in increasing the lysine and tryptophan content in the whole grain [16]. Maize varieties' improvement and QPM conversion programs, a multi-trait selection procedure using independent selection levels has been employed to increase grain yield, resistance to pest and diseases, accumulate modifiers and improve other important traits in which QPM germplasm is defective [4]. In QPM breeding program, protein and tryptophan analysis in germplasm is an important step [45, 46]. A broad range of the CIMMYT's maize populations have been converted to QPM. This germplasm is reported to have high potential for QPM cultivar development [47, 48]. QPM with high protein quality and grain yield could be accepted by the farmers [1]. QPM germplasm has been widely used for the development of QPM cultivars with high grain yield in African countries [16]. The important problem in QPM breeding is abiotic stresses. Water stress and soil infertility are the most important stresses that reduce maize productivity in developing countries. It affects major maize yield loss in African countries [49]. High land usage affects the soil fertility and decreases the nitrogen content in the soils [50]. Global climate change could influence the soil fertility and water holding capacity, and it also affects the maize production [51].

Worldwide, a large number of normal maize hybrids have been released and commercialized. But the QPM-based germplasm is quite narrow, and significantly small numbers of genetically diverse QPM hybrids are available. Nearly 12 QPM hybrids have been released in India, compared to greater than hundred normal maize hybrids [52]. In this context, it is necessary to develop various QPM varieties across the world. Conversion of QPM through conventional breeding takes at least 10–15 years. Conversion of elite normal maize hybrids into QPM hybrids requires lesser time, initially due to tested combining ability, heterosis and adaptability of the released hybrids [53]. Opaque-2 recessive allele introgression through conventional backcross breeding of 6–7 generations is required. Through marker-assisted advanced backcross breeding, time could be significantly reduced to two backcrosses [54, 55].

The opaque-2 mutation in maize inspired the research interest, with wishes to significantly increase the nutritional status of maize consumers in developing countries. QPM, which has high lysine and tryptophan, holds the security of improving the nutritional condition of children whose main staple food is maize. It is an alternative food for protein supplement in the diet. QPM has been an alternative to the people who are using synthetic lysine and tryptophan.

6.1 QPM genotypes for stress conditions

Under stress conditions, the quality of the QPM protein does not vary, but the modifications of endosperm and the content of the proteins vary greatly. To enhance the yield of QPMS under different stress conditions is the major constrain for the breeders. Drought stress affects on QPM yield mainly in grain-filling stage [56]. Some studies reported that the supply of selenium to the plant could reduce the negative effects of the water stress conditions and is considered as the cost-efficient approach to improve the quality and yield of maize [57]. Supplying nitrogen and sulfur results in the enhanced growth and yield of QPMs [58]. Some

QPM have the potential to resist some biotic stresses that are caused by some diseases and pests, but the development of QPMs that has resistance to pest or diseases that attack the grains got more importance. Thus the CIMMYT developed the QPM varieties that are resistant to some viruses and are distributed to the National Agricultural Research System (NARS) breeders that are present at different countries in 2002 [59]. During the breading process of QPMs, multiple genes are involved in enhancing the yield of grains, whereas nonadditive gene actions are highly involved for inheritance of the trait. QPM hybrids that are evaluated under salt-, drought- and *Striga*-affected conditions showed nonadditive gene action [60]. Different varieties of QPM genotypes that adapt to the environmental conditions of sub-Saharan Africa were developed by the CIMMYT (2005), and thus great benefits for children have been documented [61]. QPM hybrids could help the poor people for elevation of malnutrition in developing countries.

7. Conclusion

There is a need for the development of QPM hybrids in developing countries for protein energy source. All the agricultural research institutes have started this QPM improvement work. Through conventional breeding methodologies, the international maize research center research team has slowly improved the original opaque-2 problems. Marker-assisted breeding is an alternative method to improve the QPM production and productivity in the developing countries.

Acknowledgements

The author thanks the Head of the Department of Biotechnology, KL University, Guntur, Andhra Pradesh, India, for reviewing the book chapter.

Conflict of interest

The author declares that there is no conflict of interest on this book chapter.

Author details

S.R. Krishna Motukuri
KLEF University, Guntur, Andhra Pradesh, India

*Address all correspondence to: msrkrishna81@gmail.com

References

[1] Vasal SK. The quality protein maize story. Food and Nutrition Bulletin. 2000;**21**:445-450. DOI: 10.1177/156482650002100420

[2] Kumar A, Jat SL, Kumar R, Yadav OP, Campus P. Maize Production Systems for Improving Resource-Use Efficiency and Livelihood Security. New Delhi: DMR; 2013. p. 123. Available from: https://s3.amazonaws.com/academia.edu.documents/44282018/Maize_production_systems_for_improving_r20160331-17801-1o7ae6u.pdf?response-content-disposition=inline%3B%20filename%3DMaize_production_systems_for_improving_r.pdf&X-Amz-Algorithm=AWS4-HMAC-SHA256&X-Amz-Credential=AKIAIWOWYYGZ2Y53UL3A%2F20190806%2Fus-east-1%2Fs3%2Faws4_request&X-Amz-Date=20190806T084645Z&X-Amz-Expires=3600&X-Amz-SignedHeaders=host&X-Amz-Signature=5939035cb4b8292e04dc9ac5ec77d66af4be3e7f93ed3ad2e0da77d0710338bd

[3] Vasal SK, Villegas E, Bjarnason M, Gelaw B, Goertz P. Genetic modifiers and breeding strategies in developing hard endosperm opaque-2 materials. In: Pollmer WG, Phillips RH, editors. Improvement of Quality Traits of Maize for Grain and Silage Use. London: Martinus Nijhoff; 1980. pp. 37-73. Available from: https://www.cabdirect.org/cabdirect/abstract/19811605259

[4] Bjarnason M, Vasal SK. Breeding of quality protein maize (QPM). Plant Breeding Reviews. 1992;**9**:181-216 Available from: https://books.google.co.in/books?hl=en&lr=&id=W9aJNVA3C_0C&oi=fnd&pg=PA181&dq=Breeding+of+quality+protein+maize+(QPM)&ots=OJDw2JlGi8&sig=V5Hs2wgfklTUzXY5QmxlOCdyroo#v=onepage&q=Breeding%20of%20quality%20protein%20maize%20(QPM)&f=false

[5] Villegas E, Vasal SK, Bjarnason M. Quality protein maize - what is it and how was it developed. In: Mertz ET, editor. Quality Protein Maize. St Paul, Minnesota: American Society Cereal Chemistry; 1992. pp. 27-48

[6] Geevers HO, Lake JK. Development of modified opaque2 maize in South Africa. In: Mertz ET, editor. Quality Protein Maize. St Paul, Minnesota: American Society Cereal Chemistry; 1992. pp. 49-78

[7] Wheeler T, Von Braun J. Climate change impacts on global food security. Science. 2013;**341**:508-513. DOI: 10.1126/science.1239402

[8] FAO. World Food Programme, International Fund for Agricultural Development, the State of Food Insecurity in the World: Economic Growth Is Necessary but Not Sufficient to Accelerate Reduction of Hunger and Malnutrition. Rome: FAO; 2012. Available from: www.fao.org/docrep/016/i3027e/i3027e00.htm

[9] Temba MC, Njobeh PB, Adebo OA, Olugbile AO, Kayitesi E. The role of compositing cereals with legumes to alleviate protein energy malnutrition in Africa. Journal of Food Science and Technology. 2016;**51**:543-554. DOI: 10.1111/ijfs.1303

[10] Gibbon BC, Larkins BA. Molecular genetic approaches to developing quality protein maize. Trends in Genetics. 2005;**21**:227-233. DOI: 10.1016/j.tig.2005.02.009

[11] Anonymous Grain Zea Production Maps and Statistics. 2010. Available from: http://www.fas.usda.gov/psdonline/

[12] Prasanna BM, Vasal SK, Kassahun B, Singh NN. Quality protein maize.

Current Science. 2001;**81**:1308-1319. Available from: https://www.jstor.org/stable/24105845

[13] Yadav OP, Prasanna BM, Yadava P, Jat SL, Kumar D, Dhillon BS, et al. Doubling maize (Zea mays) production of India by 2025–challenges and opportunities. Indian Journal of Agricultural Sciences. 2016;**86**:427-434. Available from: https://www.researchgate.net/profile/Shankar_Jat/publication/301230943_Doubling_maize_Zea_mays_production_of_India_by_2025_-_Challenges_and_opportunities/links/570e3aa708aee328dd652f9c.pdf

[14] Badu-Apraku B, Annor B, Oyekunle M, Akinwale RO, Fakorede MAB, Talabi AO, et al. Grouping of early maturing quality protein maize inbreds based on SNP markers and combining ability under multiple environments. Field Crops Research. 2015;**183**:169-183. DOI: 10.1016/j.fcr.2015.07.015

[15] May T, Klatt K, Smith J, Castro E, Manary M, Caudill M, et al. Supplementation prevents a hallmark disturbance of kwashiorkor in weanling mice fed a maize vegetable diet: Hepatic steatosis of undernutrition. Nutrients. 2018;**10**:653. DOI: 10.3390/nu10050653

[16] Krivanek AF, De Groote H, Gunaratna NS, Diallo AO, Friesen D. Breeding and disseminating quality protein maize (QPM) for Africa. African Journal of Biotechnology. 2007;**6**:312-324. Available from: https://repository.cimmyt.org/xmlui/bitstream/handle/10883/3045/89937.pdf: https://repository.cimmyt.org/xmlui/bitstream/handle/10883/3045/89937.pdf

[17] Wu Y, Holding DR, Messing J. Gamma-zeins are essential for endosperm modification in quality protein maize. Proceedings of the National Academy of Sciences of the United States of America. 2010;**107**:12810-12815. DOI: 10.1073/pnas.1004721107

[18] Gayral M, Gaillard C, Bakan B, Dalgalarrondo M, Elmorjani K, Delluc C, et al. Transition from vitreous to floury endosperm in maize (Zea mays L.) kernels is related to protein and starch gradients. Journal of Cereal Science. 2016;**68**:148-154. DOI: 10.1016/j.jcs.2016.01.013

[19] Lending CR, Larkins BA. Changes in the zein composition of protein bodies during maize endosperm development. The Plant Cell. 1989;**1**:1011-1023. DOI: 10.1105/tpc.1.10.1011

[20] Sofi PA, Wani SA, Rather AG, Wani SH. Quality protein maize (QPM): Genetic manipulation for the nutritional fortification of maize. Journal of Plant Breeding and Crop Science. 2009;**6**:244-253. Available from: https://s3.amazonaws.com/academia.edu.documents/34119391/Wani_maize2009.pdf?response-content-disposition=inline%3B%20filename%3DReview_article_Quality_protein_maize_QPM.pdf&X-Amz-Algorithm=AWS4-HMAC-SHA256&X-Amz-Credential=AKIAIWOWYYGZ2Y53UL3A%2F20190806%2Fus-east-1%2Fs3%2Faws4_request&X-Amz-Date=20190806T092131Z&X-Amz-Expires=3600&X-Amz-SignedHeaders=host&X-Amz-Signature=c9e3a2dec8a5080d0cd07e71ed9ec59fd8c816cf0d6c763642dfd2605c8b0140

[21] Tripathy SK, Ithape DM, Maharana M, Prusty AM. Quality protein maize (QPM): Genetic basis and breeding perspective. Tropical Plant Research. 2017;**4**:145-152. DOI: 10.22271/tpr.2017.v4.i1.021

[22] Muller M, Muth R, Gallusci P, Knudsen S, Maddaloni M, Motto M, et al. Regulation of storage protein synthesis in cereal seeds: Developmental

and nutritional aspects. Journal of Plant Physiology. 1995;**45**:606-613. DOI: 10.1016/S0176-1617(11)81271-X

[23] Coleman CE, Larkins BA. The prolamins of maize. Seed Proteins. Kluwer Academic Publishers; 1999:109-139. DOI: 10.1007/978-94-011-4431-5_6

[24] Geetha KB, Lending CR, Lopes MA, Wallace JC, Larkins BA. Opaque-2 modifiers increase gamma-zein synthesis and alter its spatial distribution in maize endosperm. The Plant Cell. 1991;**3**:1207-1219. DOI: 10.1105/tpc.3.11.1207

[25] Ueda T, Waverczak W, Ward K, Sher N, Ketudat M, Schmidt RJ, et al. Mutations of the 22- and 27-kD zein promoters affect transactivation by the Opaque-2 protein. The Plant Cell. 1992;**4**:701-709. DOI: 10.1105/tpc.4.6.701

[26] Holding DR, Hunter BG, Chung T, Gibbon BC, Ford CF, Bharti AK, et al. Genetic analysis of opaque2 modifier loci in quality protein maize. Theoretical and Applied Genetics. 2008;**117**:157-170. DOI: 10.1007/s00122-008-0762-y

[27] Woo YM, Hu DW, Larkins BA, Jung R. Genomics analysis of genes expressed in maize endosperm identifies novel seed proteins and clarifies patterns of zein gene expression. The Plant Cell. 2001;**13**:2297-2317. DOI: 10.1105/tpc.010240

[28] Xu JH, Messing J. Organization of the prolamin gene family provides insight into the evolution of the maize genome and gene duplications in grass species. Proceedings of the National Academy of Sciences of the United States of America. 2008;**105**:14330-14335. DOI: 10.1073/pnas.0807026105

[29] Wu Y, Messing J. RNA interference-mediated change in protein body morphology and seed opacity through loss of different zein proteins. Plant Physiology. 2010;**153**:337-347. DOI: 10.1104/pp.110.154690

[30] Omosebi MO, Osundahunsi OF, Fagbemi TN. Effect of extrusion on protein quality, antinutritional factors, and digestibility of complementary diet from quality protein maize and soybean protein concentrate. Journal of Food Biochemistry. 2018;**42**:e12508. DOI: 10.1111/jfbc.12508

[31] Hossain F, Muthusamy V, Pandey N, Vishwakarma AK, Baveja A, Zunjare RU, et al. Marker-assisted introgression of opaque2 allele for rapid conversion of elite hybrids into quality protein maize. Journal of Genetics. 2018;**97**:287-298. DOI: 10.1007/s12041-018-0914-z

[32] Ortiz-Martinez M, Otero-Pappatheodorou JT, Serna-Saldívar SO, García-Lara S. Antioxidant activity and characterization of protein fractions and hydrolysates from normal and quality protein maize kernels. Journal of Cereal Science. 2017;**76**:85-91. DOI: 10.1016/j.jcs.2017.05.021

[33] Mertz ET, Bates LS, Nelson OE. Mutant gene that changes protein composition and increases lysine content of maize endosperm. Science. 1964;**145**:279-280. DOI: 10.1126/science.145.3629.279

[34] Bressani R. Nutritional value of high-lysine maize in humans. In: Mertz ET, editor. Quality Protein Maize. St. Paul, MN: American Assoc. Cereal Chem; 1992

[35] National Research Council. Quality Protein Maize. Washington, DC: National Academy Press; 1988

[36] Lopes MA, Larkins BA. Genetic analysis of opaque2 modifier gene activity in maize endosperm. Theoretical and Applied Genetics. 1995;**91**:274-281. DOI: 10.1007/BF00220889

[37] Wallace JC, Lopes MA, Paiva E, Larkins BA. New methods for extraction and quantitation of zeins reveal a high content of γ-zein in modified opaque-2 maize. Plant Physiology. 1990;**92**:191-196. DOI: 10.1104/pp.92.1.191

[38] Dannenhoffer JM, Bostwick DE, Larkins BA. Opaque-15, a maize mutation with properties of a defective opaque-2 modifier. Proceedings of the National Academy of Sciences of the United States of America. 1995;**92**:1931-1935. DOI: 10.1073/pnas.92.6.1931

[39] Gibbon BC, Wang X, Larkins BA. Altered starch structure is associated with endosperm modification in quality protein maize. Proceedings of the National Academy of Sciences of the United States of America. 2003;**100**:15329-15334. DOI: 10.1073/pnas.2136854100

[40] Paez AV, Helm JL, Zuber MS. Lysine content of opaque-2 maize kernels having different phenotypes. Crop Science. 1969;**9**:251-252. DOI: 10.2135/cropsci1969.0011183X000900020045x

[41] Nuss ET, Tanumihardjo SA. Quality protein maize for Africa: Closing the protein inadequacy gap in vulnerable populations. Advances in Nutrition. 2011;**2**:217-224. DOI: 10.3945/an.110.000182

[42] Lopes MA, Takasaki K, Bostwick DE, Helentjaris T, Larkins BA. Identification of two opaque2 modifier loci in quality protein maize. Molecular & General Genetics. 1995;**247**:603-613. DOI: 10.1007/BF00290352

[43] Yuan L, Dou Y, Kianian SF, Zhang C, Holding DR. Deletion mutagenesis identifies a haploinsufficient role for γ-zein in opaque2 endosperm modification. Plant Physiology. 2014;**164**:119-130. DOI: 10.1104/pp.113.230961

[44] Burr B, Burr FA. Zein synthesis in maize endosperm by polyribosomes attached to protein bodies. Proceedings of the National Academy of Sciences of the United States of America. 1976;**73**:515-519. DOI: 10.1073/pnas.73.2.515

[45] Vasal SK. Quality protein maize: Overcoming the hurdles. Journal of Crop Production. 2002;**6**:193-227. DOI: 10.1300/J144v06n01_11

[46] Vivek BS, Krivanek AF, Palacios-Rojas N, Twumasi-Afriyie S, Diallo AO. Breeding Quality Protein Maize (QPM): Protocols for Developing QPM Cultivars. Mexico, DF: CIMMYT; 2008

[47] Vasal SK, Srinivasan G, González F, Beck DL, Crossa J. Heterosis and combining ability of CIMMYT's quality protein maize germplasm: II. Subtropical. Crop Science. 1993a;**33**:51-57. DOI: 10.2135/cropsci1993.0011183X003300010007x

[48] Vasal SK, Srinivasan G, Pandey S, González F, Crossa J, Beck DL. Heterosis and combining ability of CIMMYT's quality protein maize germplasm: I. lowland tropical. Crop Science. 1993b;**33**:46-51. DOI: 10.2135/cropsci1993.0011183X003300010006x

[49] Bänziger M, Diallo AO. Progress in developing drought and N stress tolerant maize cultivars for eastern and southern Africa. In: Friesen DK, Palmer AFE, editors. Integrated Approaches to Higher Maize Productivity in the New Millennium. Proceedings of the 7th Eastern and Southern Africa Regional Maize Conference. Nairobi, Kenya: CIMMYT/KARI; 2004. pp. 189-193

[50] Bellon MR. Participatory methods in development and dissemination of new maize technologies. In: Pingali PL, editor. World Maize Facts

and Trends. MeetingWorld Maize Needs: Technological Opportunities and Priorities for the Public Sector. Mexico: CIMMYT 1999-2000; 2001. pp. 4-20. Available from: https://repository.cimmyt.org/bitstream/handle/10883/771/74558.pdf?sequence=1

[51] Bänziger M, Edmeades GO, Beck D, Bellon M. Breeding for Drought and Nitrogen Stress Tolerance in Maize: From Theory to Practice. Mexico, DF: CIMMYT; 2000 Available from: https://repository.cimmyt.org/xmlui/bitstream/handle/10883/765/68579.pdf

[52] Yadav OP, Hossain F, Karjagi CG, Kumar B, Zaidi PH, Jat SL, et al. Genetic improvement of maize in India: Retrospect and prospects. Agribiological Research. 2015;**4**:325-338. DOI: 10.1007/s40003-015-0180-8

[53] Prasanna BM, Pixley KV, Warburton M, Xie C. Molecular marker-assisted breeding for maize improvement in Asia. Molecular Breeding. 2010;**26**:339-356. DOI: 10.1007/s11032-009-9387-3

[54] Babu R, Nair SK, Kumar A, Venkatesh S, Sekhar JC, Singh NN, et al. Two generation marker aided backcrossing for rapid conversion of normal maize lines to quality protein maize (QPM). Theoretical and Applied Genetics. 2005;**111**:888-897. DOI: 10.1007/s00122-005-0011-6

[55] Muthusamy V, Hossain F, Thirunavukkarasu N, Choudhary M, Saha S, Bhat JS, et al. Development of β-carotene rich maize hybrids through marker assisted introgression of β-carotene hydroxylase allele. PLoS One. 2014;**9**:e113583. DOI: 10.1371/journal.pone.0113583

[56] Tandzi NL, S Mutengwa C, LM Ngonkeu E, Woïn N, Gracen V. Breeding for quality protein maize (QPM) varieties: A review. Agronomy. 2017;7:80. DOI: 10.3390/agronomy7040080

[57] Nawaz F, Naeem M, Ashraf MY, Tahir MN, Zulfiqar B, Salahuddin M, et al. Selenium supplementation affects physiological and biochemical processes to improve fodder yield and quality of maize (*Zea mays* L.) under water deficit conditions. Frontiers in Plant Science. 2016;7:1438. DOI: 10.3389/fpls.2016.01438

[58] Pavithra M, Reddy GP, Chandrika V, Umamahesh V. Nitrogen and sulphur nutrition for enhancing the growth and yield of quality protein maize (QPM). Journal of Pharmacognosy and Phytochemistry. 2018;7:142-144. Available from: http://www.phytojournal.com/archives/2018/vol7issue6/PartC/7-5-587-516.pdf

[59] Pixley KV. The Development and Promotion of Quality Protein Maize in Sub-Saharan Africa. Progress Report. Vol. 65. Harare, Zimbabwe: CIMMYT; 2003. Available from: https://repository.cimmyt.org/xmlui/bitstream/handle/10883/787/81090.pdf

[60] Annor B, Badu-Apraku B. Gene action controlling grain yield and other agronomic traits of extra-early quality protein maize under stress and non-stress conditions. Euphytica. 2016;**212**:213-228. DOI: 10.1007/s10681-016-1757-4

[61] Akalu G, Taffesse S, Gunaratna NS, De Groote H. The effectiveness of quality protein maize in improving the nutritional status of young children in the Ethiopian highlands. Food and Nutrition Bulletin. 2010;**31**:418-430. DOI: 10.1177/156482651003100304

Section 2

Adverse Effect of Changing Climate on Maize Production

Chapter 4

Climate Change Impacts on Sustainable Maize Production in Sub-Saharan Africa: A Review

Kelvin Mulungu and John N. Ng'ombe

Abstract

Maize (*Zea mays* L.) is one of the commonly grown grain crops and remains a source of staple food and food security for most countries in sub-Saharan Africa (SSA). But climate change threatens agricultural potential in SSA thereby risking food security especially that most maize production is rain-fed in these countries. Thus, numerous studies have examined impacts of climate change on maize production and productivity resulting in several adaption strategies being promoted to mitigate the negative effects of climate change. But to the best of our knowledge, there has not been any studies in literature that provide a review of impacts of climate change on maize production and productivity in SSA. This chapter therefore provides a review of empirical climate change impacts on maize production and its productivity in SSA. We chose SSA because most countries in SSA are underdeveloped and therefore more vulnerable to climate change effects. This is important because this review will provide an easier access of such results for both scholars and policy makers in search of empirical impacts of climate change on maize production and productivity in SSA.

Keywords: climate change, maize, smallholder farmers, sub-Saharan Africa

1. Introduction

Climate change is a real phenomenon worldwide [1] as observed in the increase in atmospheric and oceanic temperature, decreased amounts of snow and ice as well as a rise in sea level [2]. The earth's surface has been warmer in past three successive decades [2] resulting higher average temperature compared to the past centuries. The term "climate change" is defined differently among different stakeholders even though the contents are similar in context. IPCC [3] defines climate change as a change in the state of the climate that can be identified by changes in the mean and/or the variability of its properties, and that persists for an extended period, typically decades or longer. Based on the United Nations Framework Convention on Climate Change (UNFCCC), climate change refers to a change of climate that is attributed directly or indirectly to human activity that alters the composition of the global atmosphere and that is in addition to natural climate variability observed over comparable time periods [3].

Impacts of climate change vary depending on the state of development of a region. For example, IPCC [4] suggest that rising temperatures and changing precipitation rates will most likely hamper success of rain-fed agriculture in most developing countries. Africa is one of the continents that is projected to experience rising temperatures of at least 1 to 2°C and higher likelihood of extreme weather [5, 6]. Thus, the effects of climate change will more directly affect agriculture because about three-quarters of Africa's population depends on agriculture for a livelihood and Africa's agriculture is mainly rain-fed [7–11].

For sub-Saharan Africa (SSA), agriculture largely contributes to employment of the majority of the people in rural areas and significantly to the Gross Domestic Product (GDP) of most countries. Thus, a large number of people in SSA is employed in agriculture and increasing agricultural productivity is necessary to reducing poverty and food insecurity (AGRA [12]). However, the rise in temperatures and increased stochastic rainfall variations have both direct and indirect grave consequences on crop yields and agricultural productivity. While agriculture is so important to most developing economies in SSA, most agricultural sectors in SSA have performed poorly relative to other developing world regions [8]. Kotir [7] contends that over the past 50 years, agricultural productivity has been steadily declining in SSA and recorded the slowest increase across the world over and that this would only get worse with climate change. Taken together, this evidence suggests production of maize, a vital crop for many millions in SSA [13] may have its production in danger in the face of climate change.

Maize, a field crop that is one of the most cultivated crops in the world, is a staple crop for most countries in SSA [13]. While maize remains an important crop for many millions in SSA, its yields in developing countries (including SSA) are lower than in developed countries [14–16]. More importantly, maize production depends on water availability, and most of SSA's agriculture is rain-fed, which makes maize production an obvious candidate to be affected by weather shocks such as droughts—one of the negative consequences of climate change. Lobell et al. [17] suggest maize is sensitive to daytime high temperatures above 30°C and with climate change, the projected 2°C in temperatures for most parts of Africa would affect maize production, which would further lower maize productivity levels in SSA despite the increasing demand for maize.

Because climate change impacts are seemingly being felt, numerous studies have examined impacts of climate change on maize production and productivity resulting in several adaption strategies being promoted to negate the negative effects of climate change (e.g., [5, 14, 18–21]). To the best of our knowledge, there has not been any studies in literature that provide a comprehensive review of impacts of climate change on maize production and productivity in SSA. This chapter therefore provides a detailed review of climate change impacts on maize production and its productivity in SSA. We chose SSA because as mentioned earlier, most countries in SSA are underdeveloped and their agriculture is rain-fed—making them more vulnerable to climate change effects. This is important because this review will provide an easier access of such results for both scholars and policy makers that are in search of empirical impacts of climate change on maize production and productivity for relevant policy.

This rest of the chapter is organized as follows. The next section provides the main literature review of studies that have examined climate change impacts on maize production and productivity specifically in SSA. Adaptation to climate change as well as relative importance of temperature and rainfall are also discussed.

2. Literature review

2.1 Climate change and maize production

Climatic change is a result of anthropogenic greenhouse gas emissions which have been on the rise since the pre-industrial era. This has been largely driven by economic and population growth and the greenhouse gas emissions are now higher than before and the atmospheric concentrations of carbon dioxide, methane and nitrous oxide have increased [2, 7]. Besada and Sewankambo [1] argue that the 4th Assessment Report by the IPCC seemed to overlook Africa's concern about climate change. They claim the issue of climate change should not mainly be in terms of projections of carbon emissions and future environmental damages, but it is more about the links between climate change and contemporary disaster events which includes droughts, desertification, floods, and coastal storms. They further argue that these climate change-related disaster events eventually threaten lives and livelihoods and are a hindrance to economic growth and social progress of the continent of Africa.

Maize originated from Mesoamerica and is currently grown all over the world [13]. Maize can be grown between latitude 58°N and latitude 40°S and it grows best at moderate latitudes, but it can also be grown below sea level [22]. In Africa, 30% of the total area under cereal production is maize which accounts for over 30% of the total calories and protein consumed [14]. Of the total maize production in the developing world, 67% comes from low and lower middle-income countries which shows that maize plays an imperative role in the livelihoods of a good number of poor farmers [13].

Despite its importance, maize productivity in SSA with the exception of South Arica has remained quite low—only increased from about 0.9 to 1.5 tons/ha while the yield remains highly variable [14, 23]. The variation in yields is mainly due to dependence on rainfall under uncertain climatic conditions. With climate change, the yields of maize have been negatively affected in many regions [2]. Thus, even when compared to the top five maize producing countries in the world, maize yields in SSA have stagnated at less than two tons per hectare and less than 1.5 tons per hectare in Western and Southern Africa [14]. In addition, in SSA, the highest growth in maize area, yields and production from the year 1961 to 2010 has been West Africa when South Africa is excluded, and the lowest has been in Southern Africa with yields at a little over 1 tons/ha [24].

The prime reason put forward for this discrepancy in maize yields between SSA and other regions is less adaptive capacity of smallholder farmers to climatic change-related effects. Ng'ombe et al. [25] suggest the success of agriculture in SSA is hindered by the negative effects of climate change while [6] contend that less adaptive capacity of smallholder farmers in SSA coupled with their rain-fed farming systems (common in SSA) expose their vulnerability to climatic effects while. This observation corroborates [24] who suggest that the large gap in yield between countries in SSA and countries with comparable production conditions is larger when rain-fed areas are considered. The lower maize yields in SSA are more attributed to drought stress than other reasons such as low soil fertility, weeds, pests, diseases, low input availability, low input use and inappropriate seeds [14] and poor irrigation schemes or lack of efficient irrigation systems [26, 27].

While these climatic change-related effects on maize production may at first sight seem to be homogenous across SSA, maize production trends in some SSA countries like Zimbabwe and Zambia have changed perhaps as a result of shifts in agricultural policy. Zambia has in recent years recorded successive maize bumper harvests [28] while accessibility to subsidized farm inputs in Zambia have had a positive effect on technical efficiency of maize production in most of Zambia's provinces [29]. In contrast, the situation in Angola and Mozambique is different because prolonged civil strife and wars in the past have somehow depressed maize production and productivity trends [24]. However, being a highly susceptible crop to droughts, about 70–80% of maize losses in SSA are attributed to droughts and floods [11]. Depending on the weather conditions, farmers in some cases abandon who fields after planting [19].

According to Nelson et al. [30] the negative effects of climate change on crop production are more pronounced in SSA than in other parts of the world. Thus, severe and prolonged droughts, flooding and loss of arable land leading to reduced agricultural yields through such avenues as crop failure and loss of livestock [1] which provide draught power and household income is still probable. Literature indicates that as a result of climate change, there is an observed 10% decline in maize yield, 15% decline in rice yield and 34% decline in wheat yield in SSA in previous years [3]. Yield projections indicate that by the year 2020, yields from rain-fed agriculture in some African countries could be reduced by up to 50% which would to a great extent affect food security and worsen the malnutrition situation [3]. Mulungu et al. [31] show that for maize in Zambia under the worst-case scenario, maize yields will decrease by 25% driven mainly temperature increases offsetting the gains from increased rainfall. Hamududu and Ngoma [6] suggest decline in water availability in Zambia by 13% by the end of the century in 2100 at national level as a result of climate change which poses a much greater risk to field crops such as maize.

Africa's inability to cope with the physical, human and socioeconomic consequences of the extremes of climate makes it the most susceptible to climate change [1, 6, 7, 10, 23]. What also adds weight to the incumbent problem is that majority of maize agricultural producers in SSA reside in rural areas. For example, [5] point out that at least 83% of the 1.4 million smallholder households in Zambia grow maize—which is a huge number. But the rural poor are more vulnerable to these changes in climate and consequently, hunger, poverty and malnutrition levels will more likely continue to rise which means that the severity of climate change will increase keeping other factors constant [32]. Because of this evidence, there is need to diversify from maize production as dependence on maize production in most SSA countries is a worry for food and nutritional security, especially when alternative supplements for dietary diversity are limited [13].

According to the report by the [3] climate change will negatively affect the agricultural sector and the impact will vary by adaptation as well as rate of temperature. In line with temperature variation, the projection is that crop productivity will slightly increase at mid to high latitudes and will decrease at lower latitudes, more so in seasonally dry and tropical regions. The increase in crop productivity will occur at local mean temperature increases of up to 1–3°C and in some regions but will decrease at temperature beyond this magnitude. On the other hand, at lower latitudes, reduction in crop productivity is projected to decrease even at minor local temperature increments of 1–2°C. In particular, cereal productivity is highly likely to decrease more at lower latitudes and less at mid to high latitudes, though this would vary in some regions with temperature increase [3].

Although maize is usually considered as a warm season crop, it is actually more sensitive to high temperature stress as compared to other crops [20]. At higher temperatures, maize yields will reduce but at the same time production or multiplication of some weeds and pests will be encouraged [13]. At a high temperature of 35°C, maize yield reduces by 9% with a one-inch reduction in rainfall [23]. Thus, even if plant breeders have developed maize varieties that grow well under different biophysical environments [33], sound maize productivity is still under threat by climate change effects.

2.2 Adaptation to climate change

Research on maize has a very important role to play when it comes to adaptation to climate change in vulnerable areas [13]. Africa has been projected to be affected the most by climate change due to limited institutional, financial and technological capacity, adaptation to climate change will be difficult and complex [13]. It is expected that research and plant breeding will mitigate many of the detrimental effects but the negative effects of climate change are what is expected if farmers continued to plant the same varieties in the same way in the same areas. Some autonomous adaptations that will help offset some negative impacts of climate change include shifting of planting dates, modifying crop rotations or an uptake of pre-existing crop varieties [34].

To ensure food security for a growing population of SSA, it is very critical to adapt agricultural systems to climate change [20]. Important steps towards designing and implementing measures that are appropriate are to identify hotspots of climate change and understand associated socioeconomic impacts at different spatial scales [20]. Continued investment in maize productivity remains crucial to the growth of agriculture and food security even if there has been success in the past, which includes policies that favor maize production and productivity as well as development and adoption of new and improved maize seed and fertilizer [24]. For instance, the maize area covered by improved varieties in Ethiopia grew from 14% in 2004 to 40% in 2013 [35]. There is need to invest in research to produce a new generation of improved varieties that are tolerant to drought, resistant to pests, and nutrition-efficient [24]. Therefore, if appropriate actions are not put in place to reduce the negative effects of climate change, the danger of food insecurity is expected to increase [36]. To manage the current climate change and for future adaptation to these variations, there is need for maize varieties that are tolerant to drought, heat and water logging and are resistant to diseases and pests and insects, and to effectively contribute to mitigating climate change, practicing conservation agriculture and precision agriculture would be helpful [13].

2.3 Relative importance of temperature and rainfall

Even if temperature is an important factor in the year-to-year production, it is not as important as rainfall in determining agricultural production. In SSA, there has been some countries which had too much rainfall which led to severe flooding and unfavorable livelihood consequences. These countries included Burkina Faso in 2007 and 2009, Mozambique in 2000 and 2001, Ethiopia in 2006 and Ghana in 2007 and 2010 [7] and in the year 2017 Niger, Nigeria, Burkina Faso, Guinea, Mali, Sierra Leone, Ghana, and Central African Republic experienced floods that destroyed lives and the agricultural sector [37]. These rainfall-related disasters are more common in some countries. For example, Malawi has had 40 weather-related disasters between 1976 and 2009 [38]. Floods are very destructive and their impacts, which includes deaths and injuries of people and exposing people to toxic substances, are instant. Flooding is world over but the difference is the degree of the impacts which is dependent on the adaptive capacity of a country. Poor countries suffer more from the impacts of flooding as compared to developed countries which have high capacity to adapt [9]. Increases in temperature and variation in rainfall therefore make it less conducive for maize production in almost three quarters of countries in the world and results in yields declining [39].

However, the extreme opposite of too little rainfall, drought, is also a reality. Due to increased frequency of droughts, yields of grains and other crops could decrease substantially across the continent. The drought conditions could lead to maize being no longer grown in some areas [40]. In southern Africa, the 2002–2003 drought experience resulted in a food deficit with an estimation of 14 million people who were at a risk of starvation and in eastern Africa in 2005–2006 and 2009, maize fields were struck by severe droughts [13]. In the coming decades, so much droughts will be experienced in most of SSA [7]. More than 100 million people were affected by drought in Africa, for example over the period 1991–2008, Kenya was affected by drought about seven times which affected about 35 million people and Ethiopia was affected by drought about six times in 25 years (1983–2008) [9].

Climatic change impact on crop productivity greatly varies from region to region [8] and climate change will also affect crops differently, that is, crops like maize, rice, wheat, beans and potatoes will be highly affected and crops like millet may be less affected since they are able to resist high temperatures and low water levels [9]. However, smallholder farmers in developing countries are the most vulnerable and disadvantaged people as they entirely depend on rain-fed agriculture [8]. Cohn et al. [10] showed that in SSA and Latin America, a greater proportion of the variation in maize yields was associated with climate change. Hence, change in climate has potential to hinder sustainable development of nations by reducing production in yield which consequently leads to food insecurity [9]. However, SSA has a huge potential for expanding maize production. About 88 million hectares (88 M ha), excluding protected and forested areas, which has not yet been planted, is suited to maize production [24]. For as long as farmers replace seed every season, advantages in yield can be significant [24]. The adoption of improved open-pollinated varieties and hybrids was at 44% of maize area in Eastern and Southern Africa in 2006–2007 minus South Africa, and it was at 60% in West and Central Africa. This statistic was a suggestion of a significant increase in adopting improved varieties more so in West and Central Africa [24].

In the study done by Jones and Thornton, the global circulation model (GCM) postulated three major types of response of maize crop to climate change and these include (1) the productivity of the crop will decrease but to an extent that can be readily handled by breeding and agronomy. For example, in eastern Brazil, the changes in maize yield are predicted to be moderate with some pixels (plots of land) showing a slight yield advantage, (2) the maize crop benefits from climate change. For example in the Ethiopian highlands that surround Addis Ababa, the yields are predicted to increase even up to 100% at times although many of the pixels showing yield increases are adjacent to pixels where yields are predicted to decrease, sometimes drastically, (3) “maize yields decline drastically, all other things being equal, that major changes may have to be made to the agricultural system, or even human population may be displaced” [40].

According to [35], most of the results from Africa showed a projected yield reduction of up to −40% across all types of projections as well as sub regions even if there was a large difference in the impacts that were reported. However, only about 12% of the total sample from this study reported an increase in yield for maize grown in East, West and Northern Africa. Results for South Asia showed a similar negative projected impact but with the variation being wider [35]. Following [40] maize production is likely to reduce by 4.6 million tons per year to 2025 and this decrease will more than double to 11.6 million tons per year by 2055. In Latin America and Africa, the total production impacts of the likely future climate change to 2055 on smallholder rain-fed maize production are comparatively modest. Aggregated results, however, conceal variability, that is, in other areas yields will increase and areas where subsistence agriculture is the norm, yields will reduce [39].

Tesfaye et al. [20] outlined the biophysical impacts of climate change and the impact of climate change on maize production, consumption and food security. The biophysical impacts of climate change include changes in potential maize cultivation area, changes in maize yields and yield response to nitrogen levels. Under maize cultivation area, aggregating the change in land area suitable for maize production in SSA by the year 2050 shows a small change of 0.6–0.8% which conceals regional differences. By 2080, due to increasing areas suitable for maize cultivation in Eastern and Southern Africa, the cultivation area for SSA may increase by 1.3–2.5% whereas suitable maize cultivation areas in Central and Western Africa may reduce by

1.2–1.4%. And because of climate change, Sub-Saharan Africa countries that surround the Sahara Desert and the coastal areas of Angola are likely to lose areas that are suitable for maize production. Hence, some countries are likely to experience greater reduction in maize cultivation area by 2050 and 2080 and other countries are likely to experience an increase in maize production areas.

Under changes in maize yields, the outputs of CERES-maize (crop estimation through resource and environment synthesis) indicated a large spatial difference in maize yields under the projected climate in 2050 and 2080 across Sub-Saharan Africa. By 2050, in some parts the change in yield may be within ±5%, some may experience a reduction in yield of between 5 and 25%, other parts may experience a reduction of more than 25% and some parts of SSA may experience an increase in yields by up to 25%. By 2080, yields are likely to reduce even further in many areas and only a few will maintain the current maize yields. Under maize yield response to nitrogen levels, even if application of nitrogen increases maize yield for both the baseline and future climate conditions, the yield response of maize to nitrogen fertilizer application was less under climate conditions than the baseline conditions. However, the impact was less with high level of nitrogen application than with low level of nitrogen application. Outputs from IMPACT (International Model for Policy Analysis of Agricultural Commodities and Trade) shows that global maize production may decrease by 40–140 million tons by 2050 depending on the GCM projections. Therefore, this reduction in the global maize production may result in a decrease in global maize consumption across SSA which may lead to a decrease in daily caloric intake that is derived from maize. Consequently, the reduction in the daily caloric intake is likely to worsen food insecurity across SSA and this may result in the number of people at risk of hunger to increase by 17–37 million people.

According to [8] the impacts of climate change in Ghana in the near future are expected to worsen especially if nothing is done about the trend. With global climatic changes, the combination of abiotic and biotic stresses are likely to increase and these are damaging to crops [13]. What have been more common are the negative impacts of climate change on crops than positive impacts ([2]) and "climate change will act as a multiplier of existing threats to food security" [7]. With high confidence, the IPCC projected increases in annual mean temperatures to be larger in the tropics and subtropics of SSA than in the mid-latitudes [2]. Furthermore, by the end of this century rainfall will become more intense and more frequent over most of the mid-latitude land masses and wet tropical regions. The fifth assessment report of the IPCC reports that hazards that are related to climate worsen other stressors which have often resulted in negative outcomes for livelihoods of the poor people. Climate-related hazards affect the lives of poor people directly and indirectly through reduced crop yields or destruction of homes and increased prices for food and a reduction in food security, respectively [2]. Depending on the level of input supply and GCM projections in SSA, yields will reduce by 6–12% and 9–20% in 2050 and 2080, respectively [20]. Moreover, these figures vary according to region and the most reduction in maize yields will be in Western and Southern Africa [20]. Even if maize yields will be negatively affected by climate change by 2050 across maize mega environments (MMEs), dry and wet lowland MMEs will experience the greatest reductions [20].

Literature shows that by the end of the twenty-first century, East Africa will be likely to lose about 40% of its maize production and a general consensus is that climate change will affect maize productivity [23]. Therefore, "the impact of climate change on global maize production may cause supply shocks in maize markets across the globe which could affect food prices and, in turn, lead to some adjustments in food production, consumption and trade patterns worldwide" [20].

3. Conclusion

Climate change potentially threatens productivity and production of maize, a field crop that depends on water availability. Literature has shown that climate change effects on maize production and productivity are serious and if proper adaptation strategies to negative effects of climate change are not followed, these impacts would deepen in the near future. Governments and international agencies need to boost efforts to minimize effects of droughts, floods or in fact ensure that climate change effects are minimized. While we believe these efforts are in place, taking a longer step at improving adaption may mitigate these negative effects. For example, more competent irrigation technologies, increased research and development of drought-tolerant maize varieties, increased adoption of climate-smart adaption strategies, and call for world leaders to reconsider the negative effects of human activities on the ecosystems are highly encouraged in literature.

Acknowledgements

Authors cordially thank Cynthia Chibebe for organizing literature that enabled this work to be produced.

Conflict of interest

The authors declare no conflict of interest.

Author details

Kelvin Mulungu[1†] and John N. Ng'ombe[2,3*]

1 Department of Agricultural and Resource Economics, Colorado State University, Fort Collins, CO, USA

2 Department of Agricultural Economics and Extension, University of Zambia, Lusaka, Zambia

3 Department of Agricultural Economics, Oklahoma State University, Stillwater, OK, USA

*Address all correspondence to: ngombe@okstate.edu

† Senior authorship is not implied.

References

[1] Besada H, Sewankambo N. Climate Change in Africa: Adaptation, Mitigation and Governance Challenges. The Center for International Governance Innovation; 2009

[2] IPCC. Climate Change 2014: Synthesis Report. Fifth Assessment Report. Geneva: Intergovernmental Panel on Climate Change, IPCC; 2014

[3] IPCC. Climate Change: Synthesis Report. Fourth Assessment Report. Valencia: Intergovernmental Panel on Climate Change, IPCC; 2007

[4] IPCC. Climate Change 2013: The Physical Science Basis. Contribution of Working Group I to the Fifth Assessment Report of the Intergovernmental Panel on Climate. Cambridge, United Kingdom/New York, NY, USA: Cambridge University Press; 2013

[5] Mulenga BP, Wineman A, Sitko NJ. Climate trends and farmers' perceptions of climate change in Zambia. Environmental Management. 2017;**59**:291-306

[6] Hamududu BH, Ngoma H. Impacts of climate change on water resources availability in Zambia: Implications for irrigation development. Environment, Development and Sustainability. 2019:1-22

[7] Kotir JH. Climate change and variability in Sub-Saharan Africa: A review of current and future trends and impacts on agriculture and food security. Environment, Development and Sustainability. 2011:587-605

[8] Tetteh EK, Opareh NO, Ampadu R, Antwi KB. Impact of Climate Change: Views and perceptions of policy makers on smallholder agriculture in Ghana. International Journal of Sciences: Basic and Applied Research (IJSBAR). 2014:79-89

[9] Gemeda DO, Sima AD. The impacts of climate change on African continent and the way forward. Journal of Ecology and the Natural Environment. 2015;**7**:256-262

[10] Cohn AS, Newton P, Gil JDB, Kuhl L, Samberg L, Ricciardi V, et al. Smallholder agriculture and climate change. Annual Review. 2017:347-375

[11] Amondo E, Simtowe F. Technology innovations, Productivity and Production Risk Effects of Adopting Drought Tolerant Maize Varieties in Rural Zambia. Vancouver: International Association of Agricultural Economics (IAAE); 2018

[12] AGRA. Alliance for a Green Revolution in Africa. Africa Agriculture Status Report: Climate Change and Smallholder Agriculture in Sub-Saharan Africa. Nairobi, Kenya: Alliance for a Green Revolution in Africa (AGRA); 2014

[13] Shiferaw B, Prasanna BM, Hellin J, Bänziger M. Crops that feed the world 6. Past successes and future challenges to the role played by maize in global food security. Food Security. 2011:307-327

[14] Cairns JE, Hellin J, Sonder K, Araus JL, MacRobert JF, Thierfelder C, Prasanna BM. Adapting maize production to climate change in sub-Saharan Africa. Food Security. 2013;**5**(3):345-360

[15] Masasi B. Impact assessment of the market systems approach for revitalisation of smallholder irrigation schemes in Zimbabwe: Case study of Mutema Irrigation Scheme [MSc. thesis]. Delft, Netherlands: UNESCO-IHE. Retrieved from: https://un-ihe.on.worldcat.org/oclc/909258665

[16] Ng'ombe JN, Brorsen BW, Raun WR, Dhillon JS. Economics of the Greenseeder hand planter. Agrosystems, Geosciences & Environment. 2019;**2**:1

[17] Lobell DB, Bänziger M, Magorokosho C, Vivek B. Nonlinear heat effects on African maize as evidenced by historical yield trials. Nature Climate Change. 2011;**1**:42-45

[18] Müller C, Cramer W, Hare WL, Lotze-Campen H. Climate change risks for African agriculture. Proceedings of the National Academy of Sciences. 2011;**108**:4313-4315

[19] Mulungu K, Tembo G. Effects of weather variability on crop abandonment. Sustainability. 2018;**7**:2858-2870

[20] Tesfaye K, Gbegbelegbe S, Cairns JE, Shiferaw B, Prasanna BM, Sonder K, et al. Maize systems under climate change in sub-Saharan Africa: Potential impacts on production and food security. International Journal of Climate Change Strategies and Management. 2015:247-271

[21] Amadu FO, Miller DC, McNamara PE. Agroforestry as a pathway to agricultural yield impacts in climate-smart agriculture investments: Evidence from southern Malawi. Ecological Economics. 2020;**167**:106443

[22] Leff B, Ramankutty N, Foley JA. Geographic distribution of major crops across the world. Global Biogeochemical Cycles. 2004;**18**:1

[23] Adhikari U, Nejadhashemi AP, Woznicki SA. Climate change and eastern Africa: A review of impact on major crops. Food and Energy Security. 2015;**4**:110-132

[24] Smale M, Byerlee D, Jayne TS. Maize Revolutions in Sub-Saharan Africa. Policy Research Working Paper 5659. Nairobi: World Bank; 2011

[25] Ng'ombe JN, Kalinda TH, Tembo G. Does adoption of conservation farming practices result in increased crop revenue? Evidence from Zambia. Agrekon. 2017;**56**:205-221

[26] Ngoma H, Hamududu B, Hangoma SP, Hichaambwa M, Kabaghe C. Irrigation Development for Climate Resilience in Zambia: The Known Knowns and Known Unknowns. Working Paper 130. Lusaka: Indaba Agricultural Research Institute (IAPRI); 2017

[27] Masasi B, Ng'ombe JN. Does a market systems approach revitalize smallholder irrigation schemes? Evidence from Zimbabwe. Sustainable Agriculture Research. 2019;**8**:36-45. DOI: 10.5539/sar.v8n2p36

[28] Chapoto A, Chisanga B, Kuteya A, Kabwe S. Bumper Harvests a Curse or a Blessing for Zambia: Lessons from the 2014/15 Maize Marketing Season (No. 1093-2016-87955). 2015

[29] Ng'ombe JN. Technical efficiency of smallholder maize production in Zambia: A stochastic meta-frontier approach. Agrekon. 2017;**56**:347-365

[30] Nelson GC, Rosegrant MW, Koo J, Robertson R, Sulser T, Zhu T, et al. Impact on Agriculture and Costs of Adaptation. Food Policy Report. Washington, DC: International Food Policy Research Institute (IFPRI);

[31] Mulungu K, Tembo G, Bett H, Ngoma H. Climate change and crop yields in Zambia: Correlative historical impacts and future projections. Under Review

[32] Masipa TS. The impact of climate change on food security in South Africa: Current realities and challenges ahead. Jàmbá: Journal of Disaster Risk Studies. 2017;**9**:1-7

[33] Banziger M, Diallo AO. Progress in developing drought and N stress tolerant maize cultivars for Eastern and Southern Africa. In: Integrated Approaches to Higher Maize Productivity in the New Millennium. Proceedings of the 7th Eastern and

Southern Africa Regional Maize Conference, CIMMYT/KARI; Nairobi, Kenya; 2004. pp. 189-194

[34] Knox J, Hess T, Deccache A, Wheeler T. Climate change impacts on crop productivity in Africa and South Asia. Environmental Research Letters. 2012;7:034032

[35] Abate T, Shiferaw B, Menkir A, Wegary D, Kebede Y, Tesfaye K, et al. Factors that transformed maize productivity in Ethiopia. Food Security. 2015;7:965-981

[36] Khanal U, Wilson C, Lee B, Hoang V. Do climate change adaptations practices improve technical efficiency of smallholder farmers? Evidence from Nepal. Climatic Change. 2018;**147**:507-521

[37] UNOCHA. West and Central Africa: 2017 Flood Impact. 2017. Available from: https://www.unocha.org/ [Accessed: 30 June 2019]

[38] Pauw K, Thurlow J, Bachu M, Van-Seventer DE. The economic costs of extreme weather events: A hydrometeorological CGE analysis for Malawi. Environment and Development Economics. 2011;**16**:177-198

[39] Jones PG, Thornton PK. The potential impacts of climate change on maize production in Africa and Latin America in 2055. Global Environmental Change. 2003;**13**:51-59

[40] Ching L. Climate Change Implications for Agriculture in Sub-Saharan Africa. Rome: Food and Agriculture Organisation; 2010

Chapter 5

Amelioration of Drought Tolerance in Maize Using Rice Husk Biochar

Md. Abdul Mannan and Mariam Akter Shashi

Abstract

This chapter discussed on mitigating effects of rice husk biochar to the drought effect on maize ("BARI Hybrid Bhutta 9"). Four doses of rice husk biochar at 0, 5, 10, and 20 t/ha were applied in soil before sowing of seed. Drought treatments were maintained at 60% of field capacity and 40% of field capacity by watering every alternate day, and 80% of field capacity (control) was also maintained as well-watered treatment. Plant growth and physiological parameters were studied at 6th leaf, 10th leaf, 14th leaf, tasseling stage, cob initiation stage, and maturity stage, and yield contributing parameters were studied after harvest. Soil physical and chemical properties were studied before sowing and after harvest of crop. Drought stress reduced plant morphological growth and affected physiology and yield of maize. Mitigation of drought stress in maize was well associated with the reduction of proline content, maintenance of water-related traits, exudation rate and enhanced chlorophyll content and SPAD value, as well as dry matter production. Rice husk biochar improved the growth and yield performance of maize under drought condition. Biochar application at 20 t/ha was the best treatment to improving drought tolerance in maize.

Keywords: amelioration, drought, tolerance, maize, biochar

1. Introduction

The world population is increasing and is projected to rise by more than 1 billion by 2030 and over 2.4 billion by 2050 [1]. Therefore, to feed the increasing population, agricultural food production must be increased by 70% by 2050 [2]. In the event of growing concerns of uncertainties in climatic conditions, the abiotic stresses have become the major threat to agriculture production worldwide. Drought is one of the most important abiotic stresses which affect crop growth and yield. In Bangladesh drought is a major threat to agricultural production. As maize is usually a winter condition and due to low rainfall, the growth of maize and yield of maize are severely affected by drought stress [3]. Under drought stress, plant photosynthesis can significantly decrease, consequently reducing the amount and energy of metabolites [4] required for the proper development of both the above- and belowground biomass [5]. In severe water shortage conditions, the roots will shrink and in the leaves induced deposition. In drought conditions, reduced water potential and increased cell content of ABA regulate the metabolism of cells.

Increase in substances such as proline can be one of the major molecular responses to drought stress [6]. Drought stress-induced free radicals cause lipid peroxidation and membrane deterioration in plants [7].

Maize is the third most important cereal crops in Bangladesh, after rice and wheat. It can be cultivated year round. The crop is high yielding and rich in nutrient and has diversified uses. The demand of maize in Bangladesh is primarily from the commercial feed processing industry. This industry is the driving force of maize sector, using 80% of its aggregate maize production (excluding imports), and statistically, the poultry sector (a significant representative of feed industry) is growing at an average rate of 23% per year [8].Therefore, production of maize needs to be increased. However maize production is severely affected by drought stress. Water absorption, imbibition, and metabolic enzymatic activation are hindered under limited water availability which reduces the maize grain germination. Root and shoot elongations are parameters of seedling growth, and these are subjected to reduction by drought stress. At seedling stage in maize, reduction in shoot elongation is more than root elongation under drought stress [9]. Application of biochar is such technology which can mitigate adverse effects of drought stress on maize.

Biochar is charcoal formed from the thermal decomposition of biomass in a low- or zero-oxygen environment and at high temperatures (<700°C), and biochar production and application in soils has a very high potential for the expansion of sustainable agricultural systems and also for global climate change mitigation [10]. Experimental evidence so far shows that incorporation of biochar to soil enhanced soil water-holding capacity, improved soil water permeability, and improved saturated hydraulic conductivity (SHC) [11], modification in soil bulk density [12], and modified aggregate stability [13]. Biochar has the potential to increase the availability of plant nutrient [14]. Furthermore, research has found that biochar improves crop productivity and mitigates drought, salinity, acidity, and toxic metal stresses that are commonly associated with plant stress [15]. Biochar application increases growth and biomass of drought-stressed plants as well as increased photosynthesis [16].

Therefore, the objectives of this manuscript are to know the effects of rice husk biochar to mitigate drought effects on the growth, physiology, and yield of maize at drought conditions.

2. Mitigating effects of biochar on drought stress in maize

2.1 Effect of rice husk biochar on plant height of maize at vegetative stages under drought stress

Plant height differences of maize at vegetative stages indicated that plant height varied due to different doses of biochar under drought conditions (**Table 1**).

At the sixth leaf stage, under control condition (80% of FC), 60% of field capacity, and 40% of field capacity, highest plant heights of maize were 44.8, 43.8, and 42.2 cm, respectively, when biochar was applied at 20 t/ha, and lowest plant heights of maize were 39.4, 39.1, and 38.7 cm, respectively, when no biochar was applied. At the 10th leaf stage, under control condition (80% of FC), 60% of field capacity, and 40% of field capacity, highest plant heights of maize were 95.4, 93.0, and1.2 cm, respectively, when biochar was applied at 20 t/ha, and lowest plant heights of maize were 90.4, 89.5, and 80.2 cm, respectively, when no biochar was applied. At the 14th leaf stage, under control condition (80% of FC), 60% of field capacity, and 40% of field capacity, highest plant heights of maize were 169.3,

Biochar doses (t/ha)	6th leaf stage (cm)			10th leaf stage (cm)			14th leaf stage (cm)		
	Control	60% of FC	40% of FC	Control	60% of FC	40% of FC	Control	60% of FC	40% of FC
0	39.4de	39.1e	38.7f	90.4b	89.5b	80.2b	150.60d	139.0f	134.3f
5	42.2a–e	40.9b-e	40.2 c–e	91.2ab	90.4b	90.3b	156.6c	145.3e	136.3f
10	42.8a–c	41.9a–e	41.2b–e	93.9ab	91.4ab	90.7b	164.0b	151.3d	138.3f
20	44.8a	43.8ab	42.0a–e	95.4a	93.0ab	91.2ab	169.3a	154.3cd	145.0e
CV (%)		4.2			3.1			2.0	

Figure having similar letter did not vary significantly.

Table 1.
Effect of rice husk biochar on plant height of maize at vegetative stages under drought conditions.

154.3, and 145.0 cm, respectively, when biochar was applied at 20 t/ha, and lowest plant heights of maize were 150.6, 139.0, and 134.3 cm, respectively, when no biochar was applied. So it is clear that plant height is affected by drought conditions and application of rice husk biochar mitigated the effect of drought condition by increasing plant height. Similar result was reported in maize by [17]. Biochar promoted plant height of maize under drought conditions [18]. By affecting cell turgidity, drought impaired plant height [19]. Application of biochar can increase soil water-holding capacity which increased tissue water status and ultimately increased plant height [20].

2.2 Effect of rice husk biochar on plant height of maize at reproductive stages under drought stress

Plant height differences of maize at reproductive stages indicated that plant height varied due to different doses of biochar under drought conditions (**Table 2**).

At tasseling stage, under control condition (80% of FC), 60% of field capacity, and 40% of field capacity, highest plant heights of maize were 190.0, 184.3, and 165.6 cm, respectively, when biochar was applied at 20 t/ha, and lowest plant heights of maize were 164.0, 161.6, and 136.6 cm, respectively, when no biochar was applied. At cob initiation stage, under control condition (80% of FC), 60% of field capacity, and 40% of field capacity, highest plant heights of maize were 195.6, 190.3, and 169.0 cm, respectively, when biochar was applied at 20 t/ha, and lowest plant heights of maize were 174.3, 170.0, and 141.3 cm, respectively, when no biochar was applied. At maturity stage, under control condition (80% of FC), 60%

Biochar doses (t/ha)	Tasseling stage (cm)			Cob initiation stage (cm)			Maturity stage (cm)		
	Control	60% of FC	40% of FC	Control	60% of FC	40% of FC	Control	60% of FC	40% of FC
0	164.0cd	161.6 d	136.6 f	174.3cd	170.0d	141.3 f	175.3c	173.0c	154.0 e
5	172.6bc	172.0 bc	139.3 f	175.6cd	174.6cd	145.3 f	180.6bc	178.3bc	156.6de
10	174.3b	174.0 b	151.3 e	186.6b	182.6bc	157.6 e	186.6b	185.6b	163.0 d
20	190.0a	184.3 a	165.6bcd	195.6a	190.3ab	169.0 d	202.3a	195.6a	173.3 c
CV (%)		3.5			2.9			2.9	

Figure having similar letter did not vary significantly.

Table 2.
Effect of rice husk biochar on plant height in maize at reproductive stages under drought conditions.

of field capacity, and 40% of field capacity, highest plant heights of maize were 202.3, 195.6, and 173.3 cm, respectively, when biochar was applied at 20 t/ha, and lowest plant heights of maize were 175.3, 173.0, and 154.0 cm, respectively, when no biochar was applied. Drought conditions affected plant height, and biochar application increased plant height under drought conditions. Similar result was reported in maize by [21]. Addition of biochar improved plant height [22]. In rice, drought stress during the vegetative stage greatly reduced the plant height; [23] and [24] found that biochar increased the plant height of maize.

2.3 Effect of rice husk biochar on days to flowering of maize under drought stress

Under drought conditions plant growth as well as days to flowering of maize was affected. Days to flowering of maize varied appreciably with different doses of biochar under drought conditions (**Figure 1**).

Under control condition (80% of FC), 60% of field capacity, and 40% of field capacity, lowest days to flowering of maize were 52, 57, and 61 days, respectively, when biochar was applied at 20 t/ha, and highest days to flowering of maize were 60, 62, and 63 days, respectively, when no biochar was applied. Drought stress affected plant physiological process and biochar helps to maintain physiological activities thereby flowering of plants, improved plant growth and influenced days to flowering. [25] observed that the mung bean plants grown in soil amended with 8.5% and 15.75% wood biochar started flowering, pod filling, and maturing 6 to 7 days earlier than those grown in unamended soil.

2.4 Effect of rice husk biochar on days to maturity of maize under drought stress imposition

Plants try to avoid drought conditions by completing their life cycle within the short times. Biochar helped to reduce the effects of drought stress on crops. Days to

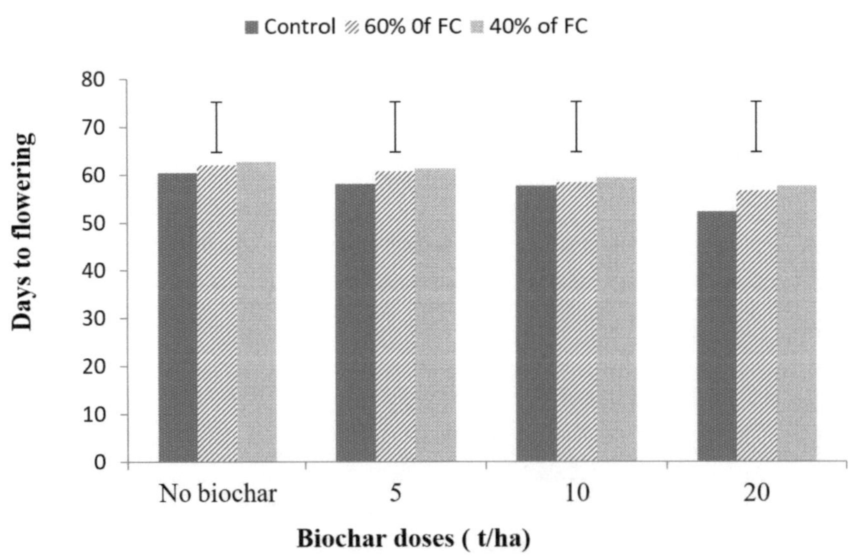

Figure 1.
Effect of rice husk biochar on days to flowering of maize under drought conditions. Bar indicates LSD at 5% level of significance.

maturity of maize plant were varied significantly at different doses of biochar under drought conditions (**Figure 2**).

Under control condition (80% of FC), 60% of field capacity, and 40% of field capacity, highest days to maturity of maize were 136, 135, and 133 days, respectively, when biochar was applied at 20 t/ha, and lowest days to flowering of maize were 131, 130, and 128 days, respectively, when no biochar was applied. Application of biochar increased the water-holding capacity of silty sand under maize cultivation in pots; [26] and [27] reported that biochar helped in maintaining normal physiological functions including maturity of wheat under saline conditions. [28] observed that biochar application increased tomato growth and life cycle under saline conditions.

2.5 Effect of rice husk biochar on relative water content (RWC), water uptake capacity (WUC), and water saturation deficit (WSD) in maize under drought stress

Relative water content of maize plant was reduced significantly at drought stress conditions because of low water content of soil. Application of rice husk biochar at different doses helped to increase water-holding capacity of soil under drought conditions and thereby increased relative water content of maize plant (**Table 3**). Under control condition (80% of FC), 60% of field capacity, and 40% of field capacity, highest RWC of maize were 83.37, 79.86, and 78.32%, respectively, when biochar was applied at 20 t/ha, and lowest RWC of maize were 66.93, 63.75, and 62.25%, respectively, when no biochar was applied.

Water saturation deficit of maize plant was increased significantly at drought stress conditions, and it is varied with different doses of biochar under drought conditions (**Table 3**).Under control condition (80% of FC), 60% of field capacity, and 40% of field capacity, lowest WSD of maize were 16.6, 20.1, and 21.1%, respectively, when biochar was applied at 20 t/ha, and highest WSD of maize were 33.0, 36.2, and 37.7%, respectively, when no biochar was applied.

Water uptake capacity of maize plant was increased significantly under drought stress because soil contained low moisture to be uptaken by plant. WUC depended

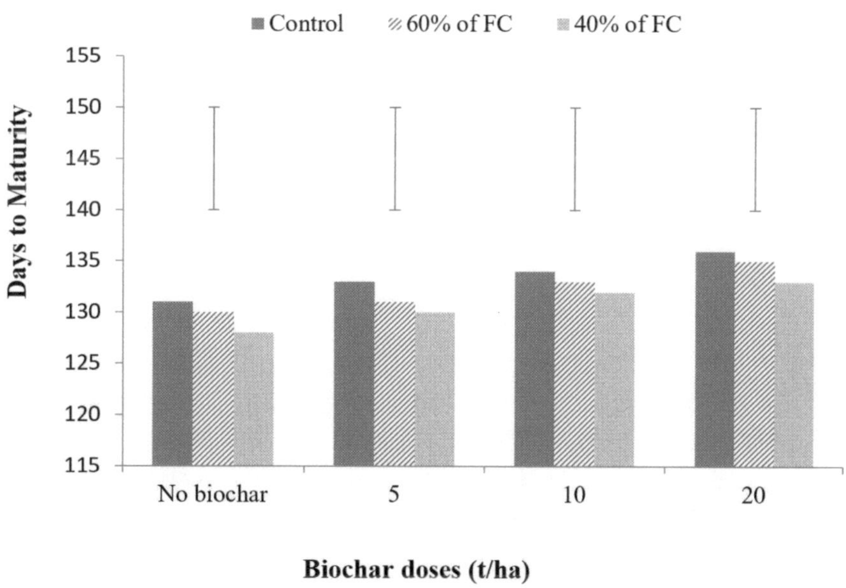

Figure 2.
Effect of rice husk biochar on days to maturity of maize under drought conditions. Bar indicates LSD at 5% level of significance.

Biochar doses (t/ha)	Relative water content (%)			Water saturation deficit (%)			Water uptake capacity		
	Control	60% of FC	40% of FC	Control	60% of FC	40% of FC	Control	60% of FC	40% of FC
0	66.9bc	63.7bc	62.2c	33.1a–c	36.2ab	37.7 a	1.9ab	1.9a	2.0a
5	71.2bc	70.2a–c	66.4bc	28.8a–c	29.7a–c	33.5a–c	1.8a–d	1.8a–c	1.9a
10	76.8bc	75.7a–c	72.8a–c	23.1a–c	24.3a–c	27.1a–c	1.7a–d	1.7a–d	1.8a–d
0	83.3a	79.8ab	78.3bc	16.6c	20.1bc	21.1a–c	1.5d	1.5cd	1.6b–d
CV (%)		14.1			36.8			10.7	

Figure having similar letter did not vary significantly.

Table 3.
Effect of rice husk biochar on RWC, WSD, and WUC of maize under drought conditions.

on water-holding capacity of soil, and it was varied with different doses of biochar under drought condition (**Table 3**). Under control condition (80% of FC), 60% of field capacity, and 40% of field capacity, lowest WUC of maize were 1.5, 1.5, and 1.6, respectively, when biochar was applied at 20 t/ha, and highest WUC of maize were 1.9, 1.9, and 2.0, respectively, when no biochar was applied. [29] reported biochar increased water-holding capacity. [30] found that biochar increased RWC and water use efficiency of drought-stressed tomato plants. [31] also reported that biochar increased tissue water status of maize in sandy soil.

2.6 Effect of rice husk biochar on exudation rate of maize under drought stress

Exudation rate of maize plant was reduced significantly at drought conditions. Exudation rate depends on available water in soil to be uptaken by the plant. Exudation rate of maize varied due to different doses of biochar under drought conditions (**Figure 3**). Under control condition (80% of FC), 60% of field capacity, and 40% of field capacity, highest exudation rates of maize were 2.3, 1.5, and

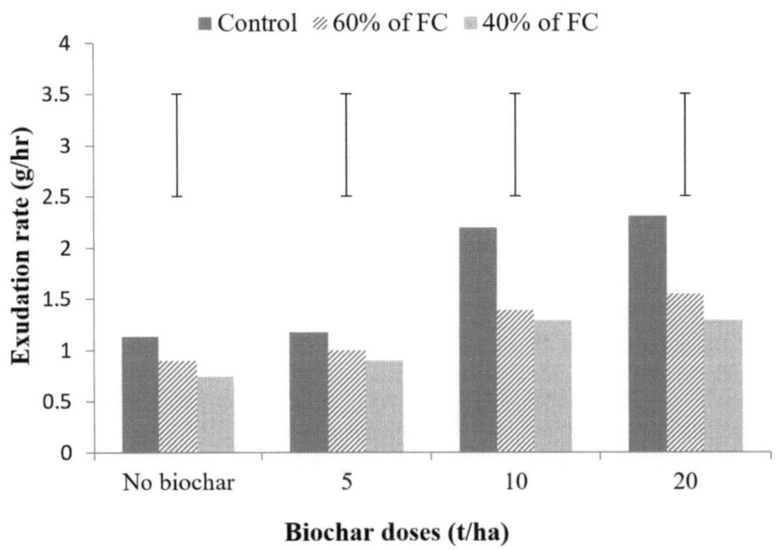

Figure 3.
Effect of rice husk biochar on exudation rate of maize under drought conditions. Bar indicates LSD at 5% level of significance.

1.5 g/hr., respectively, when biochar was applied at 20 t/ha, and lowest exudation rates of maize were 1.1, 1.0, and 0.7 g/hr., respectively, when no biochar was applied. Similar result was observed by [32]. [33] found biochar application increased water retention capacity of soil. [34] reported application of biochar increased water-holding capacity of field-grown wheat and exudation rate.

2.7 Effect of rice husk biochar on chlorophyll contents of maize under drought stress

Chlorophyll content of maize leaf was reduced significantly at drought stress conditions. Chlorophyll a content varied significantly with different doses of biochar under drought conditions (**Table 4**).

Under control condition highest chlorophyll a (1.4 mg/g) was found when biochar was applied at 20 t/ha, and it was lowest (1.2 mg/g) when no biochar was applied. Under 60% of field capacity, highest chlorophyll a (1.4 mg/g) was found when biochar was applied at 20 t/ha, and it was lowest (1.1 mg/g) when no biochar was applied. Under 40% of field capacity, highest chlorophyll a was observed when plant was treated with biochar at 20 t/ha (1.3 mg/g), and it was lowest (1.1 mg/g) when no biochar was applied. Chlorophyll b increased with the application of biochar under drought stress conditions, although it was insignificant (**Table 4**). Under control condition highest chlorophyll b (1.1 mg/g) was found when biochar was applied at 20 t/ha, and it was lowest (0.9 mg/g) when no biochar was applied. Under 60% of field capacity, highest total chlorophyll b (1.0 mg/g) was found when biochar was applied at 20 t/ha, and it was lowest (0.9 mg/g) when no biochar was applied. Under 40% of field capacity, highest chlorophyll b was observed when plant was treated with biochar at 20 t/ha (1.0 mg/g), and it was lowest (0.9 mg/g) when no biochar was applied. Under control condition highest total chlorophyll (2.0 mg/g) was found when biochar was applied at 20 t/ha, and it was lowest (1.5 mg/g) when no biochar was applied. Under 60% of field capacity, highest total chlorophyll (1.7 mg/g) was found when biochar was applied at 20 t/ha, and it was lowest (1.4 mg/g) when no biochar was applied. Under 40% of field capacity, highest total chlorophyll was observed when plant was treated with biochar at 20 t/ha (1.6 mg/g), and it was lowest (1.3 mg/g) when no biochar was applied. [39] marked reduction in chlorophylls in wheat cultivars subjected to water stress. [35] reported that biochar increased chlorophyll content in milk thistle under drought conditions.

Biochar doses (t/ha)	Chlorophyll a (mg/g fresh weight)			Chlorophyll b (mg/g fresh weight)			Total chlorophyll (mg/g fresh weight)		
	Control	60% of FC	40% of FC	Control	60% of FC	40% of FC	Control	60% of FC	40% of FC
0	1.2c–e	1.1de	1.1e	0.9a	0.9a	0.9a	1.5ab	1.4ab	1.3b
5	1.2c	1.2de	1.2c–e	1.0a	0.9a	0.9a	1.5ab	1.5ab	1.4ab
10	1.2c	1.2cd	1.2c–e	1.0a	1.0a	0.9a	1.9ab	1.5ab	1.5ab
20	1.4a	1.4ab	1.3bc	1.1a	1.0a	1.0a	2.0a	1.7ab	1.6ab
CV (%)		6.1			3.3			2.3	

Figure having similar letter did not vary significantly.

Table 4.
Effect of rice husk biochar on chlorophyll content in maize under drought conditions.

2.8 Effect of rice husk biochar on SPAD value of maize at vegetative stages under drought stress

At vegetative stage SPAD value of maize plant was reduced significantly at drought stress conditions. SPAD value varied with different doses of biochar under drought conditions (**Table 5**).

At the 6th leaf stage of maize after under control condition, highest SPAD value (30.7) was found when biochar was applied at 20 t ha^{-1}, and it was lowest (25.3) when no biochar was applied. Under 60% of field capacity, highest SPAD value (30.5) was found when biochar was applied at 20 t/ha, and it was lowest (23.7) when no biochar was applied. Under 40% of field capacity, highest SPAD value (29.5) was found when biochar was applied at 20 t/ha, and it was lowest (20.4) when no biochar was applied. At the 10th leaf stage of maize after under control condition, highest SPAD value (33.3) was found when biochar was applied at 20 t/ha, and it was lowest (29.3) when no biochar was applied. Under 60% of field capacity, highest SPAD value (30.2) was found when biochar was applied at 20 t/ha, and it was lowest (29.3) when no biochar was applied. Under 40% of field capacity, highest SPAD value (29.8) was found when biochar was applied at 20 t/ha and at 5 t/ha (29.4), and it was lowest (29.0) when no biochar was applied. At the 14th leaf stage of maize after under control condition, highest SPAD value (35.3) was found when biochar was applied at 20 t/ha, and it was lowest (29.5) when no biochar was applied. Under 60% of field capacity, highest SPAD value (32.0) was found when biochar was applied at 20 t/ha, and it was lowest (27.7) when no biochar was applied. Under 40% of field capacity, highest SPAD value (31.8) was found when biochar was applied at 20 t/ha, and it was lowest (27.2) when no biochar was applied. It indicates that the longer the exposure to drought stress, the higher the decreases of the SPAD value. The decrease of SPAD reading under drought conditions is reported by [36]. [37] showed that biochar may alleviate water stress in plants and increased SPAD value.

2.9 Effect of rice husk biochar on SPAD value of maize at reproductive stages under drought stress

SPAD value of maize plant was reduced significantly at drought conditions, and reduction was higher at 40% field capacity than 60% of field capacity at tasseling stage and cob initiation stage (**Table 6**).

Biochar doses (t/ha)	6th leaf stage			10th leaf stage			14th leaf stage		
	Control	60% of FC	40% of FC	Control	60% of FC	40% of FC	Control	60% of FC	40% of FC
0	25.3c–e	23.7e	20.4f	30.4cd	29.3ef	29.0f	29.5de	27.7e	27.2e
5	27.5bc	25.2de	24.4de	30.7c	29.9de	29.4ef	33.0a–c	29.9b–e	28.9de
10	29.5ab	26.4cd	26.2cd	32.0b	30.0cd	29.7d–f	33.3ab	31.1bd	29.7c–e
20	30.7a	30.5a	29.5ab	33.3a	30.2cd	29.8d–f	35.3a	32.0a–d	31.8b–d
CV (%)		5.0			1.6			6.6	

Figure having similar letter did not vary significantly.

Table 5.
Effect of rice husk biochar on SPAD value in maize at vegetative stages under drought conditions.

Biochar doses (t/ha)	Tasseling stage			Cob initiation stage		
	Control	60% of FC	40% of FC	Control	60% of FC	40% of FC
0	30.2bc	28.0cd	27.8d	28.2b–d	27.6cd	27.1d
5	30.6b	29.4c–d	29.2b–d	29.8a–c	29.3a–d	29.2a–d
10	30.9b	29.8b–d	29.7b–d	30.7ab	29.5a–d	29.5a–d
20	33.5a	31.2b	30.7b	31.3a	31.0a	30.7ab
CV (%)		4.4			5.0	

Figure having similar letter did not vary significantly.

Table 6.
Effect of rice husk biochar on SPAD value in maize at reproductive stages under drought conditions.

When biochar was applied at different doses, SPAD value was increased. At tasseling stage of maize under control condition (80% of FC), 60% of field capacity, and 40% of field capacity, highest SPAD values were 33.5, 31.2, and 30.7, respectively, when biochar was applied at 20 t/ha, and lowest SPAD values were 30.2, 28.0, and 27.8, respectively, when no biochar was applied. At cob initiation stage of maize under control condition (80% of FC), 60% of field capacity, and 40% of field capacity, highest SPAD values were 31.3, 31.0, and 30.7, respectively, when biochar was applied at 20 t/ha, and lowest SPAD values were 28.2, 27.6, and 27.1, respectively, when no biochar was applied. Similar result was reported by Mannan et al. (2016) in soybean plant under salinity stress due to poultry litter biochar. With increasing drought stress levels, SPAD readings were decreased [38]. [39] reported biochar increased soil moisture level and maize yield.

2.10 Effect of rice husk biochar on proline of maize under drought stress

Proline is a kind of stress protein. Proline accumulation under stress condition occurred because the Calvin cycle of photosynthesis is affected by drought; as a result N content could not be properly metabolized. In drought soil biochar increases photosynthesis and proper metabolism of N content. Proline content of maize varied significantly with different doses of biochar under drought conditions (**Figure 4**).

Under control condition (80% of FC), 60% of field capacity, and 40% of field capacity, lowest proline contents were 1.1, 1.1, and 3.2 μ mole/g, respectively, when biochar was applied at 20 t/ha, and highest proline contents were 1.8, 2.9, and 6.1 μ mole/g, respectively, when no biochar was applied. [40] reported biochar decreased proline content in plants. [41] marked drought stress caused overproduction of proline content. [42] also reported biochar increased photosynthesis in grape leaves.

2.11 Effect of rice husk biochar on dry weight of cob sheath, leaf, and stem of maize under drought stress

A major effect of drought is reduction in photosynthesis, which is associated with reduction in food production and ultimately reduced dry weight of plant parts. Dry weight of cob sheath, leaf, and stem of maize is greatly affected by drought conditions. Application of rice husk biochar increased dry matter of cob sheath, leaf, and stem of maize under drought conditions. Dry weight of cob sheath, leaf, and stem of maize varied significantly with different doses of biochar under drought conditions (**Table 7**).

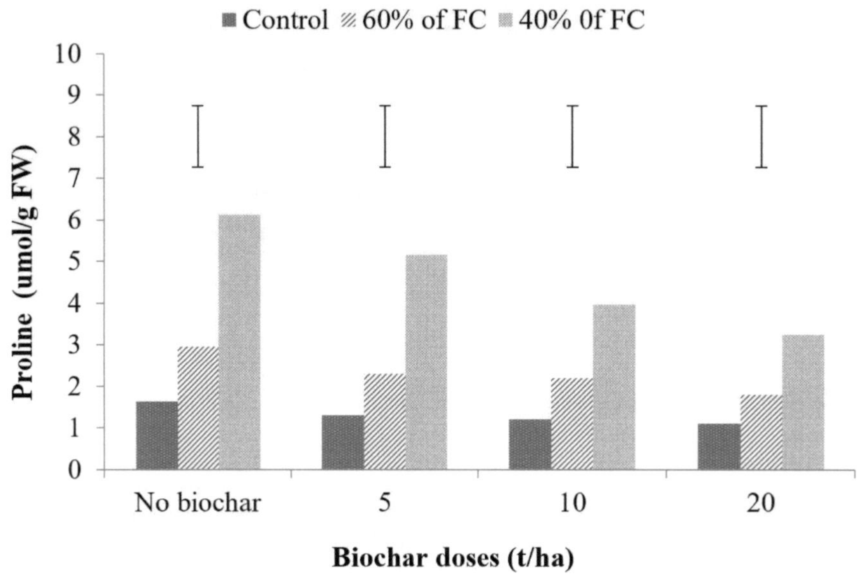

Figure 4.
Effect of rice husk biochar on proline content of maize under drought conditions. Bar indicates LSD at 5% level of significance.

Biochar doses (t/ha)	Cob sheath (g/plant)			Leaf (g/plant)			Stem (g/plant)		
	Control	60% of FC	40% of FC	Control	60% of FC	40% of FC	Control	60% of FC	40% of FC
0	12.8bc	11.7bc	10.3c	37.2d–g	35.2fg	34.4g	24.8ab	21.4ab	19.0b
5	13.1bc	12.6bc	11.4bc	39.8a–d	36.4e–g	36.2e–g	25.4ab	22.3ab	20.6ab
10	14.5bc	14.2bc	12.9bc	40.3a–c	39.4a–d	37.7c–f	26.8a	25.2ab	21.3ab
20	19.7a	15.5ab	14.8abc	42.0a	41.5ab	38.6b–e	27.2a	26.2a	22.7ab
CV (%)		21.9			4.6			17.3	

Figure having similar letter did not vary significantly.

Table 7.
Effect of rice husk biochar on dry weight of cob sheath, leaf and stem of maize under drought conditions.

The highest dry weight of stem were 27.28 g, 26.25 g and 22.75 g in control, 60% of field capacity and 40% of field capacity, respectively, when biochar was applied at 20 t/ha, and lowest dry weights of cob sheath were 12.8, 11.7, and 10.3 g, respectively, when no biochar was applied. Under control condition (80% of FC), 60% of field capacity, and 40% of field capacity, highest dry weights of leaf were 42.0, 41.5, and 38.6 g, respectively, when biochar was applied at 20 t/ha, and lowest dry weights of leaf were.

Table 7. Effect of rice husk biochar on dry weight of cob sheath, leaf, and stem of maize under drought conditions, 37.2, 35.2, and 34.4 g, respectively, when no biochar was applied. Under control condition (80%of FC), 60% of field capacity, and 40% of field capacity, highest dry weights of stem were 27.2, 26.2, and 22.7 g, respectively, when biochar was applied at 20 t/ha, and lowest dry weights of stem were 24.8, 21.8, and 19.0 g, respectively, when no biochar was applied. [43] found drought stress reduced dry weight of plant parts by affecting photosynthesis. [44] reported that application of biochar increased dry weight of field-grown wheat.

2.12 Effect of rice husk biochar on shoot, root, and total dry weight of maize under drought stress

In drought stress shoot dry weight of maize reduced, but root dry weight increased, because under drought conditions for searching water, root growth increased, thereby increasing dry weight of root. Application of rice husk biochar reduced the effects of drought. The dry weight of root and shoot varied significantly with the application of biochar under drought conditions (**Table 8**).

Under control condition (80% of FC), 60% of field capacity, and 40% of field capacity, highest dry weights of shoot were 84.1, 83.1, and 75.9 g, respectively, when biochar was applied at 20 t/ha, and lowest dry weights of shoot were 75.4, 68.4, and 63.8 g, respectively, when no biochar was applied. Under control condition (80% of FC), 60% of field capacity, and 40% of field capacity, lowest dry weights of root were 12.4, 15.6, and 16.8 g, respectively, when biochar was applied at 20 t/ha, and highest dry weights of root were 17.5, 26.8, and 27.3 g, respectively, when no biochar was applied. Total dry weight of maize plant was reduced at drought stress conditions, but reduction was not significant. When biochar is applied at different doses under drought conditions, total dry weight increased (**Table 8**). Under control condition (80% of FC), 60% of field capacity, and 40% of field capacity, highest total dry weights were 98.8, 97.0, and 93.9 g, respectively, when biochar was applied at 20 t/ha, and lowest total dry weights were 93.1, 91.1, and 89.3 g, respectively, when no biochar was applied. [45] found that root dry weight increased, while shoot dry weight decreased under drought conditions. [46] marked shoot dry weight increased under drought conditions due to application of biochar.

2.13 Effect of rice husk biochar on number of cob, length of cob, and diameter of cob of maize under drought stress

The number of cob was one per plant, and there is no significant difference among numbers of cob per plant under drought stress condition with different biochar doses (**Table 9**).

Drought affected growth of maize. Length of cob of maize was reduced under drought conditions. When biochar was applied at different doses, the cob length was increased under drought conditions (**Table 9**). Under control condition highest cob length (17.6 cm) was found when biochar was applied at 20 t/ha, and it was lowest (15.9 cm) when no biochar was applied. Under 60% of field capacity, highest

Biochar doses (t/ha)	Shoot dry weight (g/plant)			Root dry weight (g/plant)			Total dry weight (g/plant)		
	Control	60% of FC	40% of FC	Control	60% of FC	40% of FC	Control	60% of FC	40% of FC
0	75.4a–d	68.4cd	63.8d	17.5a–c	26.8a	27.3a	93.1a	91.1a	89.3a
5	77.8a–c	71.5b–d	68.2cd	16.7a–c	17.7a–c	24.4ab	95.2a	94.6a	92.6a
10	81.2ab	79.3a–c	72.9a–d	15.7bc	16.3a–c	21.7a–c	95.6a	95.5a	92.8a
20	84.1a	83.1ab	75.9a–c	12.4c	15.6bc	16.8a–c	98.8a	97.0a	93.9a
CV (%)		9.3			34.4			8.7	

Figure having similar letter did not vary significantly.

Table 8.
Effect of rice husk biochar on shoot, root, and total dry weight of maize under drought conditions.

Biochar doses (t/ha)	Number of cob			Length of cob (cm)			Diameter of cob (cm)		
	Control	60% of FC	40% of FC	Control	60% of FC	40% of FC	Control	60% of FC	40% of FC
0	1.0a	1.0a	1.0a	15.9a–c	13.2bc	12.1c	3.5a–c	3.2c	3.1c
5	1.0a	1.0a	1.0a	16.5ab	14.7a–c	14.6a–c	3.6a–c	3.3bc	3.2c
10	1.0a	1.0a	1.0a	17.2ab	15.1a–c	15.0a–c	3.8ab	3.5a–c	3.3a–c
20	1.0a	1.0a	1.0a	17.6a	15.3aa–c	15.3a–c	3.9a	3.6a–c	3.5a–c
CV (%)		0.0			15.7			2.15	

Figure having similar letter did not vary significantly.

Table 9.
Effect of rice husk biochar on number of cob, length of cob, and diameter of cob of maize under drought conditions.

cob length (15.3 cm) was found when biochar was applied at 20 t/ha, and it was lowest (13.2 cm) when no biochar was applied. Under 40% of field capacity, highest total cob length (15.3 cm) was found when biochar was applied at 20 t/ha, and it was lowest (12.1 cm) when no biochar was applied. Cob diameter of maize was reduced under drought stress conditions, and reduction was higher at 40% of field capacity than at 60% of field capacity. Biochar application increased cob diameter under drought conditions (**Table 9**). Under control condition highest cob diameter (17.6 cm) was found when biochar was applied at 20 t/ha, and it was lowest (15.9 cm) when no biochar was applied. Under 60% of field capacity, highest cob diameter (15.3 cm) was found when biochar was applied at 20 t/ha, and it was lowest (13.2 cm) when no biochar was applied. Under 40% of field capacity, highest total cob diameter (15.3 cm) was found when biochar was applied at 20 t/ha, and it was lowest (12.1 cm) when no biochar was applied. [47] reported biochar increased yield of lettuce. Reductions in plant yield have been reported in snap bean by [48]. [49] observed biochar application increased maize yield in semiarid conditions.

2.14 Effect of rice husk biochar on number of seed/cob, 100 grain weight (g), and grain yield (g) of maize under drought stress conditions

Drought stress affected anthesis, grain filling of maize associated with reduction of number seed/cob, 100 grain weight, and ultimately grain yield. Decrease of photosynthesis under drought conditions also affected grain yield. Application of biochar increased photosynthesis efficiency, anthesis, and grain filling, thereby increasing yield of maize. The number of seed per cob, 100 grain weight, and grain yield varied significantly with biochar doses under drought conditions (**Table 10**).

Under control condition (80%of FC), 60% of field capacity, and 40% of field capacity, highest numbers of seed per cob were 353.0, 335.0, and 334.6, respectively, when biochar was applied at 20 t/ha, and lowest seeds per cob were 163.0, 147.3, and 139.0, respectively, when no biochar was applied. Under control condition highest 100 grain weight (27.7 g) was found when biochar was applied at 20 t/ha, and it was lowest (21.8 g) when no biochar was applied. Under 60% of field capacity, highest 100 grain weight (26.5 g) was found when biochar was applied at 20 t/ha, and it was lowest (20.7 g) when no biochar was applied. Under 40% of field capacity, highest 100 grain weight (25.0 g) was found when biochar was applied at 20 t/ha, and it was lowest (20.0 g) when no biochar was applied. Under control condition (80% of FC), 60% of field capacity, and 40% of field capacity, highest

Biochar doses (t/ha)	Number of seed /cob			100 grain weight (g)			Grain yield (g/plant)		
	Control	60% of FC	40% of FC	Control	60% of FC	40% of FC	Control	60% of FC	40% of FC
0	163.0bcd	147.3cd	139.0d	21.8a–c	20.7bc	20.0c	40.7cd	35.9cd	27.8d
5	273.0a–d	244.0a–d	164.3b–d	23.4a–c	21.7a–c	21.4a–c	58.6a–d	57.5a–d	34.9cd
10	300.0ab	297.0a–c	288.3a–d	26.8ab	23.0a–c	21.5a–c	79.5ab	68.9a–c	61.0a–d
20	353.0a	335.0a	334.6a	27.7a	26.5a–c	25.0a–c	96.7a	89.7ab	84.5ab
CV (%)		35.5			16.70			37.40	

Figure having similar letter did not vary significantly.

Table 10.
Effect of rice husk biochar on the number of seed/cob, 100 grain wt. (g), and grain yield (g) of maize under drought conditions.

grain yields were 96.7, 89.7, and 84.5 g/plant, respectively, when biochar was applied at 20 t/ha, and lowest grain yields were 40.7, 35.9, and 27.8 g/plant, respectively, when no biochar was applied. Similar result was reported by [50]. [51] observed water stress reduced yield of triticale. [52] reported biochar increased pod yield of soybean under saline conditions.

2.15 Effect of rice husk biochar on N, P, and K in soil under drought stress

Under drought conditions biological activities as well as nutrients in soil are greatly affected. As a result macronutrients such as N, P, and K are reduced. Application of rice husk biochar showed positive effects on total nitrogen content and P and K under stress and nonstressed conditions (**Table 11**).

The initial total N was 0.17%, and after crop harvest under control condition, the highest total N (0.14%) was found when biochar was applied at 20 t/ha; it was lowest (0.10%) when no biochar was applied. Under 60% of field capacity, highest total N (0.13%) was found when biochar was applied at 20 t/ha, and it was lowest (0.10%) when no biochar was applied. Under 40% of field capacity, highest total N (0.11%) was found when biochar was applied at 20 t/ha, and it was lowest (0.09%) when no biochar was applied. The initial P was 7.24 ppm, and after harvest under control condition (80% of FC), 60% of field capacity, and 40% of field capacity,

Before sowing	Total N (%)			P (ppm)			K (meq/100 g soil)		
	0.172			7.24			0.169		
	After harvest								
Biochar doses (t/ha)	Control	60% of FC	40% of FC	Control	60% of FC	40% of FC	Control	60% of FC	40% of FC
0	0.10a	0.10a	0.09a	7.49bc	7.48bc	7.44c	0.17a	0.17a	0.17a
5	0.11a	0.11a	0.10a	7.96bc	7.74bc	7.61bc	0.17a	0.17a	0.17a
10	0.12a	0.11a	0.11a	9.13a	7.98bc	7.64bc	0.18a	0.17a	0.17a
20	0.14a	0.13a	0.11a	9.18a	8.00b	7.96bc	0.18a	0.18a	0.17a
CV (%)		7.0			4.0			1.5	

Figure having similar letter did not vary significantly.

Table 11.
Effect of rice husk biochar on N, P, and K in soil under drought conditions.

highest P were 9.18, 8.00, and 7.96 ppm, respectively, when biochar was applied at 20 t/ha, and lowest P were 7.49, 7.48, and 7.44 ppm, respectively, when no biochar was applied. The initial K was 0.16 meq/100 g soil, and after crop harvest under control condition (80% of FC), 60% of field capacity, and 40% of field capacity, highest K were 0.18 meq/100 g soil, 0.18 meq/100 g soil, and 0.17 meq/100 g soil, respectively, when biochar was applied at 20 t/ha, and lowest K were 0.17 meq/ 100 g soil, 0.17 meq/100 g soil, and 0.17 meq/100 g soil, respectively, when no biochar was applied. [53] reported biochar increased plant available nutrient in soil. [54] reported drought reduced N, P, and K levels in soil. [55] observed that the addition of biochar to soils increased soil phosphorus (P), soil potassium (K), and total soil nitrogen (N).

2.16 Effect of rice husk biochar on Zn, pH, and organic carbon (OC) in soil under drought stress

Drought stress adversely affected soil chemical properties such as Zn, pH, and OC. Application of rice husk biochar increased Zn, pH, and OC in soil. Zn and soil pH varied significantly with different doses of rice husk biochar under drought conditions, but OC varied insignificantly (**Table 12**).The initial Zn content was 17.4 meq/100 g soil, and after crop harvest under control condition (80% of FC), 60% of field capacity, and 40% of field capacity, highest Zn were 17.4 meq/100 g soil, 15.3 meq/100 g soil, and 14.9 meq/100 g soil, respectively, when biochar was applied at 20 t/ha, and lowest Zn were 13.9 meq/100 g soil, 13.2 meq/100 g soil, and 12.6 meq/100 g soil, respectively, when no biochar was applied. The initial pH was 6.1, and after crop harvest under control condition (80% of FC), 60% of field capacity, and 40% of field capacity, highest pH were 7.0, 6.9, and 6.7, respectively, when biochar was applied at 20 t/ha, and lowest pH were 6.7, 6.7, and 6.6, respectively, when no biochar was applied. The initial OC was 1.4%, and after crop harvest under control condition (80% of FC), 60% of field capacity, and 40% field capacity, highest OC were 0.7, 0.7, and 0.6%, respectively, when biochar was applied at 20 t/ha, and lowest OC were 0.54, 0.53, and 0.52%, respectively, when no biochar was applied. Similar result was reported by [56]. [57] marked biochar improved soil chemical properties of saline soil and biochar increased organic carbon. [58] found that biochar increased soil pH, thus reducing lime requirements.

Before sowing	**Zn (meq/100 g soil)**			**pH**			**OC (%)**		
	17.49			**6.18**			**1.45**		
	After harvest								
Biochar doses (t/ha)	**Control**	**60% of FC**	**40% of FC**	**Control**	**60% of FC**	**40% of FC**	**Control**	**60% of FC**	**40% of FC**
0	13.9b–e	13.2de	12.6e	6.7ab	6.7b	6.6b	0.5a	0.5a	0.5a
5	14.3b–e	14.0b–e	13.2de	6.7ab	6.7ab	6.7ab	0.5a	0.5a	0.5a
10	15.7ab	14.8b–d	13.9c–e	6.9a	6.7ab	6.7ab	0.6a	0.6a	0.59a
20	17.4a	15.3bc	14.9b–d	7.0a	6.9a	6.7ab	0.7a	0.7a	0.6a
CV (%)		7.4			2.9			6.8	

Figure having similar letter did not vary significantly.

Table 12.
Effect of rice husk biochar on Zn, pH, and organic carbon in soil under drought conditions.

3. Conclusions

Application of rice husk biochar increased plant height, days to maturity, total dry weight, chlorophyll content, plant water relations, SPAD value, exudation rate and reduced proline content, and days to flowering of maize under drought conditions. In maize plant drought stress tolerance ameliorate rice husk biochar and increased cob diameter, cob length, 100 grain weight of cob, seed /cob and finally maize yield at drought conditions.

Acknowledgements

We are grateful to the University Grants Commission (UGC), Government of Bangladesh, for funding the work.

Conflict of interest

There is no conflict of interest.

Author details

Md. Abdul Mannan* and Mariam Akter Shashi
Department of Agronomy, Bangabandhu Sheikh Mujibur Rahman Agricultural University, Gazipur, Bangladesh

*Address all correspondence to: mannanbsmrau@yahoo.com

References

[1] United Nations (UN). Department of Economic and Social Affairs and Population Division. World Population Prospects. The 2015 Revision, Key Findings and Advance Tables. Working Paper, No. ESA/P/WP; 2015. p. 241

[2] Wani SH, Sah SK. Biotechnology and abiotic stress tolerance in rice. Journal of Rice Research. 2014;**2**:1000-1105. DOI: 10. 4172/jrr.1000105

[3] WPSA, Report. Commercial broiler growth. Report on Maize value Chain in Northern Char Areas in Bangladesh. Prepared by OXFAM. Dhaka, Bangladesh; 2013

[4] Kulkarni M, Phalke S. Evaluating variability of root size system and its constitutive traits in hot pepper (Capsicum annum L.) under water stress. Science of Horticulture. 2009; **120**:159-166

[5] Dias PC, Araujo WL, Moraes GA, Barros RS, DaMatta FM. Morphological and physiological responses of two coffee progenies to soil water availability. Journal of Plant Physiology. 2007;**164**:1639-1647

[6] Matysik J, Alia B, Mohanty P. Molecular mechanisms of quenching of reactive oxygen species by proline under stress in plants. Current Science. 2002;**82**:525-532

[7] Nair A, Abraham TK, Jaya DS. Studies on the changes in lipid peroxidation and antioxidants in drought stress induced cowpea (Vigna unguiculata L.) varieties. Journal of Environmental Biology. 2008;**29**(5): 689-691

[8] WPSA, Report. Commercial broiler growth. Report on Maize value Chain in Northern Char Areas in Bangladesh. Prepared by OXFAM. Dhaka, Bangladesh; 2013

[9] Khodarahmpour. Drought stress in maize (*Zea mays* L.) effects, resistance mechanism and global achievements. 2011. VIII. 74p. Available from: http://www.springer.com/978-3-319-25440-1

[10] Lehmann J, Joseph S. Biochar for environmental management: Introduction. In: Lehmann J, Joseph S, editors. Biochar for Environmental Management: Science and Technology. London: Eathscan; 2009. pp. 1-12

[11] Asai H, Samson BK, Stephan HM, Songyikhangsuthor K, Homma K, Kiyono Y, et al. Biochar amendment techniques for upland rice production in northern Laos 1. Soil physical properties, leaf SPAD and grain yield. Field Crops Research. 2009;**111**(1):81-84

[12] Laird DA, Brown RC, Amonette JE, Lehmann J. Review of the pyrolysis platform force producing bio-oil and biochar. Biofuels, Bioproducts and Biorefining. 2010;**3**:547-562

[13] Peng X, Ye LL, Wang CH, Zhou H, Sun B. Temperature and duration dependent rice straw-derived biochar: Characteristics and its effects on soil properties of an ultisol in southern China. Soil and Tillage Research. 2011; **112**:159-166

[14] Lehmann J, Solomon D, Kinyangi J, Dathe L, Wirick S, Jacobsen C. Spatial complexity of soil organic matter forms at nanometer scales. Nature and Geological Science. 2008;**1**:238-242

[15] Thomas SC, Frye S, Gale N, Garmon M, Launchbury R, Winsborough C. Biochar mitigates negative effects of salt additions on two herbaceous plant species. Journal of Environmental Management. 2013;**129**: 62-68

[16] Rizwan M, Qayyum M, Ali S, Ok Y, Ibrahim M, Riaz M, et al. Biochar soil

amendment on alleviation of drought and salt stress in plants: A critical review. Environmental Science and Pollution Response. 2017. DOI: 10.1007/s11356-017-8904-x

[17] Abukari A. Effect of rice husk biochar on maize productivity in the guinea savannah zone of Ghana, Doctoral dissertation. Department of Agroforestry, Kwame Nkrumah University of Science and Technology; 2014

[18] Lehmann J, Rillig MC, Thies J, Masiello CA, Hockaday WC, Crowley D. Biochar effects on soil biota—A review. Soil Biology and Biochemistry. 2011;**43**: 1812-1836

[19] Hussain M, Malik MA, Farooq M, Ashraf MY, Cheema MA. Improving drought tolerance by exogenous application of glycinebetaine and salicylic acid in sunflower. Journal of Agronomy and Crop Science. 2008;**194**: 193-199

[20] Kim HS, Kim KR, Yang JE, Ok YS, Owens G, Nehls T, et al. Effect of biochar on reclaimed tidal land soil properties andmaize (Zea maysL.) response. Chemosphere. 2016;**142**:153-159

[21] Batool A, Taj S, Rashid A, Khalid A, Qadeer S, Ghufran MA. Potential of soil amendments (biochar and gypsum) in increasing water use efficiency of Abelmoschus esculentus L. Moench. Frontiers in Plant Science. 2015;**6**:1-13

[22] Hardy S, Dunst G, Glaser B. No effect level of co- composted biochar on plant growth and soil properties in a greenhouse experiment. Agronomy. 2014;**4**(1):34-51

[23] Manikavelu A, Nadarajan N, Ganesh SK, Gnanamalar RP, Babu RC. Drought tolerance in rice: Morphological and molecular genetic consideration. Plant Growth Regulator. 2006;**50**:121-138

[24] Haider G, Koyro HW, Azam F, Steffens D, Müller C, Kammann C. Biochar but not humic acid product amendment affected maize yields via improving plant-soil moisture relations. Plant and Soil. 2015;**395**:141-157

[25] Carnaje NP, Amparado RF Jr, Malaluan RM. Amending acidic soil with bamboo (Bambusa blumeana) biochar: Effect on mung bean (Vigna radiata) growth rate and yield. AES Bioflux. 2015;**7**(1):109-123

[26] Haider G, Koyro HW, Azam F, Steffens D, Müller C, Kammann C. Biochar but not humic acid product amendment affected maize yields via improving plant-soil moisture relations. Plant and Soil. 2015;**395**:141-157

[27] Akhtar SS, Andersen MN, Liu F. Residual effects of biochar on improving growth, physiology and yield of wheat under salt stress. Agriculture and Water Management. 2015;**158**:61-68

[28] Usman ARA, Al-Wabel MI, Abdulaziz AH, Mahmoud WA, El-Naggar AH, Ahmad M, et al. Conocarpus biochar induces changes in soil nutrient availability and tomato growth under saline irrigation. Pedosphere. 2016;**26**:27-38

[29] Igalavithana AD, Ok YS, Niazi NK, Rizwan M, Al-Wabel MI, Usman AR, et al. Effect of corn residue biochar on the hydraulic properties of sandy loam soil. Sustain. 2017;**9**:1-10

[30] Akhtar SS, Li G, Andersen MN, Liu F. Biochar enhances yield and quality of tomato under reduce irrigation. Agriculture Water Management. 2014;**138**:37-44

[31] Uzoma KC, Inoue M, Andry H, Fujimaki H, Zahoor A, Nishihara E. Effect of cow manure biochar on maize productivity under sandy soil condition. Soil Use and Management. 2011;**27**: 205-212

[32] Lee SS, Shah HS, Awad YM, Kumar S, Ok YS. Synergy effects of biochar and polyacrylamide on plants growth and soil erosion control. Environment and Earth Science. 2015;**4**:2463-2473

[33] Abel S, Peters A, Trinks S, Schonsky H, Facklam M, Wessolek G. Impact of biochar and hydrochar addition on water retention and water repellency of sandy soil. Geoderma. 2013;**202**:183-191

[34] Olmo M, Alburquerque JA, Barrón V, del Campillo MC, Gallardo A, Fuentes M, et al. Wheat growth and yield responses to biochar addition under Mediterranean climate conditions. Biological Fertilizer for Soil. 2014;**50**:1177-1187

[35] Afshar RK, Hashemi M, DaCosta M, Spargo J, Sadeghpour A. Biochar application and drought stress effects on physiological characteristics of *Silybum marianum*. Community Soil Science Plant Analysis. 2016;**47**:743-752

[36] Netto AT, Campostrini E, de Oliveira JG, Bressan-Smith RE. Photosynthetic pigments, nitrogen, chlorophyll a fluorescence and SPAD-502 readings in coffee leaves. Scientia Horticulturae. 2005;**104**:199-209

[37] Liu C, Liu F, Ravnskov S, Rubæk GH, Sun Z, Andersen MN. Impact of wood biochar and its interactions with mycorrhizal fungi, phosphorus fertilization and irrigation strategies on potato growth. Journal of Agronomy and Crop Science. 2017;**203**: 131-145

[38] Netto AT, Campostrini E, de Oliveira JG, Bressan-Smith RE. Photosynthetic pigments, nitrogen, chlorophyll a fluorescence and SPAD-502 readings in coffee leaves. Scientia Horticulturae. 2005;**104**:199-209

[39] Haider G, Steffens D, Moser G, Müller C, Kammann CI. Biochar reduced nitrate leaching and improved soil moisture content without yield improvements in a four-year field study. Agriculture, Ecosystems and Environment. 2017;**237**:80-94

[40] Xiao Q, Zhu LX, Shen YF, LiS Q. Sensitivity of soil water retention and availability to biochar addition in rain fed semi-arid farm land during a three-year field experiment. Field Crops Response. 2016;**196**:284-293

[41] Taiz L, Zeiger E. Plant Physiology. 4th ed. Massachusetts: Sinauer Associates Incorporation with Publishers; 2006

[42] Baronti S, Vaccari FP, Miglietta F, Calzolari C, Lugato E, Orlandini S, et al. Impact of biochar application on plant water relations in Vitis vinifera (L.). European Journal of Agronomy. 2014; **53**:38-44

[43] Earl H, Davis RF. Effect of drought stress on leaf and whole canopy radiation use efficiency and yield of maize. Journal of Agronomy. 2003;**95**: 688-696

[44] Olmo M, Alburquerque JA, Barrón V, del Campillo MC, Gallardo A, Fuentes M, et al. Wheat growth and yield responses to biochar addition under Mediterranean climate conditions. Biological Fertilizer for Soil. 2014;**50**:1177-1187

[45] Morizet T, Pollucsck M, Togola D. Drought tolerance in four maize varieties. Field Crop Abstracts. 1986;**39**:306

[46] Kamara A, Hawanatu SK, Mohamed SK. Effect of Rice straw biochar on soil quality and the early growth and biomass yield of two Rice varieties. Agricultural Sciences. 2015;**6**: 798-806

[47] Artiola JF, Rasmussen C, Freitas R. Effects of a biochar-amended alkaline

soil on the growth of romaine lettuce and Bermuda grass. Soil Science. 2012; **177**:561-570

[48] Lakitan B, Wolfe DB, Zobel RW. Flooding affects snap bean yield and genotypic variation in leaf gas exchange and root growth response. Journal of the American Society for Horticultural Science. 1992;**117**:711-716

[49] Foster EJ, Hansen N, Wallenstein M, Cotrufo MF. Biochar and manure amendments impact soil nutrients and microbial enzymatic activities in a semi-arid irrigated maize cropping system. Agriculture, Ecosystems and Environment. 2016;**233**: 404-414

[50] Pereira RG, Heinemann AB, Madari BE, Carvalho MTDM, Kliemann HJ, Santos APD. Transpiration response of upland rice to water deficit changed by different levels of eucalyptus biochar. Pesquisa Agropecuária Brasileira. 2012;**47**:716-721

[51] Estrada-Campuzano G, Miralles DJ, Slafer GA. Genotypic variability and response to water stress of pre- and post-anthesis phases in triticale. European Journal of Agronomy. 2008; **28**:171-177

[52] Mannan MA, Halder E, Karim MA, Ahmed JU. Alleviation of adverse effects of drought stress on soybean (Glycine max L.) by using poultry litter biochar. Bangladesh Agronomy Journal. 2016; **19**(2):61-69

[53] Egamberdieva D, Reckling M, Wirth S. Biochar-based Bradyrhizobium inoculum improves growth of lupin (Lupinus angustifolius L.) under drought stress. European Journal of Soil Biology. 2017;**78**:38-42

[54] Garg BK. Nutrient uptake and management under drought: Nutrient-moisture interaction. Agriculture. 2003; **27**:1-8

[55] Biederman LA, Harpole WS. Biochar and its effects on plant productivity and nutrient cycling: A meta-analysis. GCB Bioenergy. 2013; **5**(2):202-214

[56] Rizwan M, Ali S, Qayyum MF, Ibrahim M, Rehman MZ, Abbas T, et al. Mechanisms of biochar-mediated alleviation of toxicity of trace elements in plants: A critical review. Environment Science and Pollution Response. 2016;**23**:2230-2248

[57] Wu Y, Xu G, Shao HB. Furfural and its biochar improve thegeneral properties of a saline soil. Solid Earth. 2014;**5**:665-671

[58] Van Zwieten L, Kimber S, Morris S, Chan KY, Downie A, Rust J, et al. Effects of biochar from slow pyrolysis of paper mill waste on agronomic performance and soil fertility. Plant and Soil. 2010;**327**:235-246

Section 3

Improved Management Approaches for the Sustainability of Maize Production

Chapter 6

Improved Technologies for Higher Maize Production

Manpreet Jaidka, Shikha Bathla and Ramanjit Kaur

Abstract

An array of production technologies, from land preparation to harvesting, has been recommended for maize crop. Being non-tillering crop, optimum plant population can be achieved if suitable crop establishment techniques like method of sowing, sowing time, seed rate, seed treatment, crop geometry etc., are followed. Weeds can be managed well either by two hoeings 15–30 days after sowing or herbicides like atrataf 50 WP (atrazine) at 2 kg/ha on medium to heavy textured soils and 1.25 kg/ha in light soils within 10 days of sowing, using 500 litres as pre-emergence or spray 262.5 ml/acre laudis 420 SC (tembotrione) in 375 litres of water at 20 days after sowing. Integrated nutrient management strategy renders use of farm yard manure at 10–15 t/ha, Paddy straw compost at 450 kg/ha or synthetic fertilizers at 120 kg N, 60 kg P_2O_5 and 40 kg K_2O per hectare for hybrids and 80 kg N, 30 kg P_2O_5 and 20 kg K_2O per hectare for composites. Integrated pest management approach emphasizes on use of physical, chemical or biological measures for the control of insect-pests. Maize borer can be controlled by spraying coragen 18.5 SC at 75 ml using 150 litres water/ha. Drying of maize produce can be done sun drying, smoking or air drying for fetching better market price.

Keywords: maize, production technologies, crop establishment, integrated nutrient management, integrated pest management, maize drying

1. Introduction

Maize is known as the Queen of Cereals' due to its' demand and wider adaptability. It is the second most important cereal crop in the world in terms of acreage and production. Global production of Maize was about 1040 million MT in the year 2016–2017, where in USA and China contributed about 38 and 23%, respectively. In India, maize is the 3rd most important food crop after rice and wheat, where about 15 million farmers are engaged in maize cultivation [1]. In India, Andhra Pradesh ranks first in maize production followed by Karnataka with per cent share of 20.9 and 16.5, respectively [2]. It has a share of 9% in about Rs. 100 billion agriculture sector gross domestic product [3]. Maize can be cultivated successfully in loamy sand to heavy clay, well aerated, neutral pH soils. As of tropical origin, it is highly sensitive to water stagnation, so avoid the cultivation in low-lying or poor drainage fields. Furthermore, extended low temperature less 5°C severally affects the crop. Optimum range of temperature for better crop growth and yield realization is 25–35°C [4]. Being day neutral, maize crop can be cultivated throughout the year which leads to high yield levels in a short period of time. In this chapter, we are going to discuss an array of different production technologies to be followed by farmers for successful cultivation and better realization of yields. A brief outline of the chapter is given below.

2. Origin and distribution

Central America and Mexico is the primary centre of origin of maize which consists of a diversity of maize crop. Various studies reveal that maize crop was a significant crop in Mexico about 5000 years ago. USA has the largest area under maize crop followed by Brazil, China, Mexico and India. USA also stands first in terms of production followed by China. In India, Uttar Pradesh, Bihar, Rajasthan, Madhya Pradesh and Punjab are the major maize growing states. Highest acreage and production is in Uttar Pradesh while average yield/ha is recorded in Andhra Pradesh [2, 5].

3. Climatic requirements

Maize crop can grow under diverse conditions from sea level to about 3000 m altitude throughout the year in many parts of the country. In Northern India, *kharif* (monsoon) season is main growing period while in Southern India it can be grown from April to October as warm weather conditions prevail for longer period. Maize crop requires 21 and 32°C temperature for proper germination and growth with considerable moisture availability. For instance, 50–75 cm of well distributed rain is conducive for proper growth. During flowering, high temperature and low humidity damages the foliage, desiccates the pollens interferes with pollination and decreased grain formation. Maize is highly sensitive to water stagnation especially during early period of growth [2, 5].

4. Improved production technologies

4.1 Crop establishment

4.1.1 Selection of cultivar

Type of cultivar/hybrid to be grown depends on the crop season namely, spring, *Kharif* or *Rabi*. Cultivars can be proffered based on length of growing season, availability of optimum moisture regime. Depending upon above factors, cultivars can be selected as follows (**Table 1**) [6]:

4.1.2 Sowing time

Due to occurrence of diverse climatic conditions in country, planting time varies from place to place. Optimum planting time in different agro-climatic regions is described in **Table 2** [2]. The optimum time to sow the crop depends on availability of irrigation facilities. For example, if irrigation facilities are available, maize crop can be sown about 2 weeks before onset of monsoon while under rainfed conditions, crop is sown with the onset of monsoon to have optimum moisture regime so that proper plant stand can be maintained in field. In Punjab, Maize crop can be sown during all seasons at following sowing times (**Table 3**) [7, 8]:

Length of cropping season (days)	Type of cultivar
More than 100	Late maturing
90 to 100	Medium maturing
80 to 90	Early maturing

Table 1.
Choice of cultivar as per length of growing season.

Agro-climatic region	Optimum planting time
Indo-gangatic plains	15 June–15 July
North-western hills	April-early May
North-eastern hills	First fortnight of March
Peninsular region	May–June

Table 2.
Optimum planting of maize in different agro-climatic regions.

Season	Planting time
Kharif	Last week of May to last week of June
Spring	20th of Jan to 15th of Feb

Table 3.
Season wise planting time maize.

Hybrids	20–25 kg/ha
Composites	18–20 kg/ha

Table 4.
Seed rate of maize hybrids and composites.

4.1.3 Seed rate

Being a non-tillering crop it cannot compensate for the lost space if proper plant stand is not maintained under field conditions. So maintenance of 60–65,000 plants/ha is pre-requisite for realizing maximum yield. Sowing of the crop should be done 60 × 20–25 cm crop geometry. For hybrids and composites, seed rate can be used with respect to seed weight and requirement of plant population as given in **Table 4** [2, 8].

4.1.4 Seed treatment

Seed treatment plays a pivotal role in prevention of diseases and availability of nutrients to growing crop. For instance, seed treatment of maize with Bavistin or Derosal or Agrozim 50 WP (Carbendazim) @ 3 g/kg seed prevents the attack of seed and soil borne diseases in maize crop. Furthermore, treatment of seed with consortium (biofertilizer) @ 1.25 kg/ha helps in yield enhancement and improvement of soil health [2, 7, 8].

Purpose	Crop geometry
Grain crop	60 cm × 20 cm; 75 cm × 20 cm
Baby corn	30 cm × 20 cm; 60 cm × 15 cm
Fodder	30 cm × 10 cm

Table 5.
Crop geometry of maize to be followed as per requirement.

4.1.5 Crop geometry

Crop geometry has direct effect on inter and intra-plant competition in field crops. Maize crop can be planted in varied crop geometries (**Table 5**) depending upon the purpose of cultivation [2, 8]. Interculture operations like thinning, gap filling and earthing-up play critical role in performance of maize crop. Thinning needs to be performed about 10 days after germination to keep 1 plant/hill. Further, 2 earthing-ups are required in maize crop. First at 35–40 and 2nd at 60–65 days after germination [9].

4.1.6 Method of planting

Although crop establishment is a series of events that depends on interactions of seed, soil moisture, method of sowing, machinery etc. but method of planting plays an important role in establishment of crop under given set of conditions. Maize is mainly sown directly through seed by using different methods of tillage & establishment. Recently, resource conservation technologies (RCTs) namely, zero tillage, minimum tillage, surface seeding etc. had came in practice in various maize based cropping system and are cost effective and environment friendly. Following are major planting methods that vary from situation to situation.

4.1.7 Zero tillage

Maize crop can be cultivated without any primary tillage under no-till (**Figure 1**) with decreased cost of cultivation and better resource use efficiency. In this situation, maintenance of proper soil moisture at sowing and band placement of seed and fertilizers with zero-till seed-cum-fertilizer planter with furrow opener as per the soil texture and field conditions is pre-requisite. The technology is followed by large number of farmers especially under rice-maize and maize-wheat systems in peninssular and eastern India. If the field is infested with weeds, farmers can go for foliar spray of gramoxone 24 SL (paraquat) @ 1250 ml/ha about 24 hours before planting of maize crop [2, 7, 8].

4.1.8 Ridge/raised bed planting

This planting method (**Figure 2**) is considered best for cultivation during monsoon and winter seasons both under excess and limited water availability conditions. On non-uniform lands, this method is most suitable for successful cultivation of maize crop. Planting of crop needs to be done on the southern side of the east–west ridges/beds for better exposure to sunlight during winters and better crop stand. Raised bed planter having inclined plate, cupping or roller type seed dropping system should be used for planting that facilitates proper placement of seed and fertilizers in single operation for having good crop stand, higher productivity and resource use efficiency. Irrigation water can be saved to the tune of

Figure 1.
Maize crop sown under zero tillage system.

Figure 2.
Planting of maize crop on the ridges.

Figure 3.
Flat sowing of maize crop.

20–30%. Under temporary excess soil moisture/water logging due to heavy rains, the furrows will act as drainage channels and crop can be saved from excess soil moisture stress [2, 5, 7, 8].

4.1.9 Flat sowing

Maize crop can be cultivated by conventional tillage flat planting (**Figure 3**) depending upon soil type and availability of irrigation facilities. Light soils have high

Figure 4.
Maize crop establishment through transplanting system.

infiltration rate and low water holding capacity, so farmers can go for flat planting of maize crop. Under rainfed conditions, to have better moisture availability to crop for longer period, flat planting becomes better alternate. Flat planting is also beneficial when no tillage system gets infested with high weed population and chemical/manual weed control becomes non-economical [7, 8].

4.1.10 Transplanting

It is better establishment technique winter maize (**Figure 4**) in the intensive cropping system where field cannot be vacated on time, to prevent the delayed planting and crop loss due to low temperature. Under this situation, nursery of the crop is raised on a smaller portion of land and seedlings are transplanted in required field as and when they achieve certain age. For example, if the fields are to be vacated during December–January, it is advisable to go for nursery sowing 30–40 days before the transplanting. Seedlings can be transplanted in the furrows followed by light irrigation [2, 5].

4.1.11 Furrow planting

Furrow planting (**Figure 5**) of maize is recommended when crop is to be cultivated during spring season as high evaporative losses may lead to water deficit stress in flat and raised bed or ridge sowing [2, 5, 7, 8].

4.2 Water management

Water requirement of the maize crop varies from 400 to 600 mm [10]. Excess or shortage of moisture can have harmful impact on the crop growth. Proper drainage of standing water and meeting the crop needs at critical stages play a pivotal role in better crop performance. Especially for winter maize, it is advisable to keep soil wet (frequent & mild irrigation) during 15 December to 15 February to protect the crop from frost injury [3].

4.2.1 Flood irrigation

Flood method of irrigation is followed where maize crop is cultivated with flat sowing. Crop is irrigated as and when required. Generally, young seedlings, knee high stage (V8), flowering (VT) and grain 7.

Figure 5.
Crop establishment by furrow planting.

filling (GF) are critical stages and hence irrigation should be ensured at these stages [2, 7, 8].

4.2.2 Furrow irrigation

When crop is cultivated as ridge/raised bed planting, furrow irrigation is followed. Care needs to be taken at first irrigation that water should not overflow on the ridges/beds. As a thumb rule, the irrigation should be applied in furrows up to 2/3rd height of the ridges/beds. In raised bed and in limited irrigation water, the irrigation water can also be applied in alternate furrows to save irrigation water. In rainfed conditions, tied-ridges prove helpful in conserving the rainwater, increasing its availability in the root zone for longer period [2, 7, 8, 11].

4.2.3 Above ground drip irrigation

High temperature and high evaporative demand during summer season enhances the water requirement of maize crop as a result of which farmers go for a number of irrigation. To increase the water use efficiency of crop, above ground drip irrigation is recommended by Punjab Agricultural University. In this, broad beds are prepared at 1.20 m apart from centre to centre of furrow. These beds are 80 cm wide on the top and 40 cm wide furrows between beds. The beds are covered with U.V stabilized plastic film (Black) of 25 micron thickness (23 grams per m^2). Two rows of maize are planted at a spacing of 60 cm keeping plant to plant distance of 20 cm. One lateral pipe is used to irrigate two rows of maize. The drippers are spaced 30 cm apart and are operated at a discharge of 2.2 L per hour as given in **Table 6** [7, 8, 12]. Prevailing climatic regimes of an area affect the efficiency of drip irrigation system [12].

* If discharge rate is different, time of irrigation may be adjusted proportionally by the formula:

$$\text{Time of irrigation} = \frac{2.2 \times \text{Time of irrigation (min.)} *}{\text{Discharge of dripper (l/hr)}} \tag{1}$$

4.2.4 Sub-surface drip irrigation

In field experiments, sub surface drip irrigation and fertigation resulted in 18.4% higher system productivity with saving of 28.5% applied irrigation water. Sub-surface irrigation technology can be followed in maize-wheat-summer moong

Month	Timing of irrigation (min)
February	22
March	64
April	120
May	130

Table 6.
Month-wise timing of above ground drip irrigation in spring maize.

Crop	Month	Timing of irrigation (min)
Maize	July	35
	August	35
	September	50
	October	30
Wheat	December	30
	January	65
	February	70
	March	50
Summer Moong	May	60
	June	45

Table 7.
Month-wise timing of sub-surface drip irrigation in maize-wheat-summer moong cropping system.

cropping system. For this system, Place drip inline having dripper having 20 cm spacing at 20 cm depth with lateral to lateral spacing of 67.5 cm for sub surface drip irrigation in maize-wheat-summer moong cropping system. Sow one row of maize, two rows of wheat and two rows of summer moong on each drip inline during respective season. If discharge of the dripper is 2.2 L/hour, the schedule given in **Table 7** can be followed for sub-surface drip irrigation in above mentioned cropping system [7, 8, 10].

If discharge rate is different, then time of irrigation may be adjusted proportionally by the formula:

$$= (2.2 \times \text{time of irrigation (minutes)} \ast) \div \text{dripper discharge (litre/h)} \tag{2}$$

4.2.5 Partial root drying irrigation

This technique (**Figure 6**) involves alternate wetting and drying of two halves of root zone of crop plants during consecutive irrigations. The PRD technique was developed on the basis of knowledge of root-to-shoot chemical signaling (can be negative or positive) about soil conditions that regulates the shoot physiology. Alternating is essential for maintaining a constant emission of signals from the root-to-shoot, because prolonged exposure of root to drying soil may cause anatomical changes which reduce the ability of root to sense soil drying and not able to sustain the production of ABA for long time period [10]. Different methods to apply the PRD technique can be separation of root system into two parts with sheet particularly in pots, controlled alternate surface drip irrigation on half part of the

Figure 6.
Field view of partial root drying irrigation technique in maize.

root zone, controlled alternate subsurface drip irrigation on half part of the root zone or controlled alternate furrow irrigation [10].

4.3 Weed management

Maize crop is infested with grassy and broad leaf annual weeds. Among grassy, *Dactyloctenium aegypticum*, *Eleucine indica*, *Setaria glauca*, *Cyanodon dactylon*, *Cyperus rotundus*, *Sorghum helepanse*, *Bracharia rapens* are common. The broad leaf weeds are *Celosia argentia*, *Commelina bengalensis*, *Phylanthis niruri*, *Solanum nigrum*, *Amaranthus viridis*, *Trianthema partulacastrum*. Effective weed management strategies have key role in successful maize cultivation. Adoption of weed control practices during the first 6–8 weeks after planting is crucial because maize crop kept weed free for 30–45 days after planting is almost similar in yield as that kept weed free for entire crop period. The annual yield loss in maize because of weed problems is estimated to be approximately 10%. A number of weed management approached can be followed for weed management in maize crop that can be as follows [2, 7, 8]:

4.3.1 Non-chemical control: manual weeding, mulching

Non-chemical weed control measures can physical or cultural that means manual removal of weeds from the maize fields. In cultural method, Give two hoeings 15–30 days after sowing with khurpa/kasaula/wheel-hoe/triphali/tractor-drawn cultivator. Mulching is practice of keeping crop residues or plastic sheets on the soil surface within the crop rows. Mulching helps in temperature regulation, water conservation as well weed control in field crops [7, 8].

4.3.2 Chemical control

Sometimes due to continuous rains during the early stages of maize growth it becomes impossible to enter in the field. Also due to scarce availability of farm labour, the only effective way to control weeds is the use of herbicides. Spray of atrataf 50 WP (atrazine) @ 2 kg/ha on medium to heavy textured soils and 1.25 kg/ha in light soils within 10 days of sowing, using 500 L of water prove propitious in keeping weed population low in maize fields. Spray the herbicide uniformly at recommended rates to minimize residual toxicity to crops sown after maize. Alternatively, spray 262.5 ml/ha laudis 420 SC (tembotrione) in 375 L of water at 20 days after sowing provides effective control of mixed weed flora. For the control of *Cyperus rotundus* (dila/motha), apply 500 ml/ha 2,4-D amine salt 58 SL as post emergence 20–25 days after sowing in 375 L of water [2, 7, 8].

4.4 Nutrient management

4.4.1 Integrated nutrient management

Among the cereal crops, maize in general and specifically hybrids are very responsive to nutrients applied through organic or inorganic means. The rate of application depends on soil nutrient status and cropping system. For realizing required yield, the dose of applied nutrients should be as par the soil supplying capacity and crop demand. As the response of maize crop to organic manures is remarkable so integrated nutrient management (INM) is very important option in maize based systems.

- Apply 10–15 t/ha of good quality farmyard manure per hectare to the maize crop year after year [7, 8].

- Green manure the field, to be put under maize with Dhaincha/Sunhemp/Cowpea. Cowpea/Dhaincha/Sunhemp should be sown during second fortnight of April using 12/20/20 kg seed per acre, respectively. The 50 days old green manure crop should be burried and allowed to decompose for about 10 days before sowing of maize. In case, summer moong crop is grown the straw should be burried before sowing of maize [7, 8].

- Inoculate the maize seed with recommended bio-fertilizer as described earlier. For this, mix half kg packet of recommended consortium bio-fertilizer with 1 L of water and then thoroughly mix it with maize seed on clean pucca floor. Let it dry in shade and sow the seed immediately. Inoculation with bio-fertilizer should be done after treating the seed with fungicide. The seed inoculation with consortium biofertilizer increase grain yield as well as improves soil health [7, 8, 11].

- Paddy straw compost @ 450 kg/ha along with recommended dose of fertilizers can be an alternate to farm yard manure [7, 8].

- As a general recommendation, one could apply 120 kg N, 60 kg P_2O_5 and 40 kg K_2O per hectare for hybrids and 80 kg N, 30 kg P_2O_5 and 20 kg K_2O per hectare for composites. Drill one third of nitrogen and the entire quantity of phosphorous and potassium at the time of sowing. Top dress one third of nitrogen at the knee-high stage and the remaining one third at the pre tasseling stage. It may be noted that application of nitrogen fertilizer more than recommended dose is no substitute for FYM [7, 8].

- Decreased Zn availability visuals emerge on middle leaves (2nd or 3rd from tip) of plants which include white or light yellow band and reddish veins on both sides of the midrib [7, 8]. Remedial measures are described in **Table 8**:

4.4.2 Fertigation

It refers to simultaneous application of irrigation water and fertilizers by drip irrigation. By this method, FUE can go up to 80%. In drip irrigation model for spring maize, certain recommendations are made in respect to fertilizer application along with drip irrigation. For the medium fertility soils application of 200 kg of urea, 80 kg of mono ammonium phosphate (MAP) and 40 kg of muriate of potash (white)/ha is recommended. Start fertigation 12 days after sowing of maize and apply 25% of the fertilizers in four equal splits during first month on weekly basis. Rest of the fertilizer should be applied in equal splits on weekly basis upto first week of May. Furthermore,

Method of application	$ZnSO_4$ (33%)	$ZnSO_4$ (21%)
Broadcasting	16.25 kilogram/ha	25 kilogram/ha
Foliar application	1.88 + 0.94 kilogram unslaked lime	3 + 1.5 kilogram unslaked lime

Table 8.
Remedial measures for Zn deficiency in maize.

in sub-surface drip irrigation, fertilizer can be applied to maize crop when grown in maize-wheat-summer moong cropping system. For instance, Apply sub surface drip irrigation at 3 days interval for maize and summer moong with fertigation of 80% recommended dose of NPK. In maize, apply 1/5 dose of NPK at sowing and fertigate remaining P and K in 5 splits and N in 7 splits at 9 days interval starting from 15 DAS. Apply sub surface drip irrigation at 7 days interval up to mid-February and thereafter at 5 days interval to wheat with fertigation of 80% recommended dose of NPK. In wheat, apply 1/5th dose of NPK at sowing and fertigated the remaining NPK in 8 splits at 7 days interval starting from crown root initiation. In summer moong, fertigated NPK dose in 5 equal splits at 6 days interval starting from 10 DAS. Use urea, mono ammonium 119 phosphate and muriate of potash as source of N, P and K, respectively [7, 8].

4.5 Insect and pest management

4.5.1 Integrated pest management (IPM)

IPM (**Figure** 7) is highly efficient and eco-friendly strategy which includes integrated use of all possible alternates that can be biological, physical, cultural or chemical for controlling pests. Growers who are aware of the potential for pest infestation follow a four-tiered approach. The four steps include: set action thresholds, monitor and identify pests, prevention and control [11, 13].

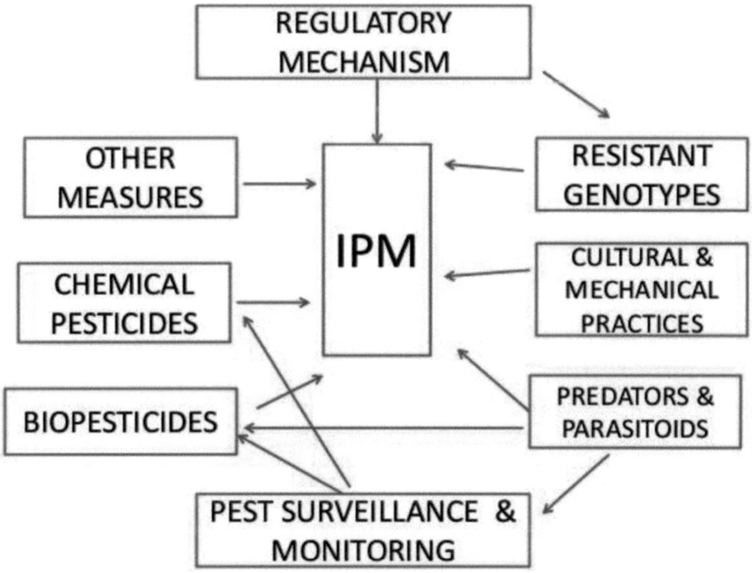

Figure 7.
Components of IPM.

- **Cultural control**: Deep summer plowing helps in destroying resting stage of pests. Inter-cropping with legume reduces borer incidence. Use of well decomposed farm yard manure termite attack. Balanced use of fertilizers.

- **Genetic management**: Use of good quality planting material from reliable source.

- **Mechanical control**: Cutting and destroying infected plants which ceases further spread. Use of pheromone traps. Set up of light traps.

- **Chemical control**: use of synthetic chemicals for the control of insect-pest and diseases [13].

4.5.2 Biological pest management

This approach encompasses use of living entities for the control of insect-pests and diseases. Living entities can be predators, herbivores or parasites along with intensive human interference. For controlling maize borer and other insects, apply bio-insecticides like Neemazal (1%) @ 300 ml/ha. The maize borer can also be managed by using tricho-cards twice having 40,000 eggs of Corcyra parasitized by *Trichogramma chilonis*. Make first release on 10 days old crop and second 1 week after first release. Cut tricho-cards into 40 equal strips and staple them uniformly on the underside of the central whorl leaves in evening hours. The tricho-cards should not be applied on rainy days [8, 11].

4.5.2.1 Major insect-pests and diseases

Maize stem borer: This insect (**Figure 8**) attacks the maize crop mainly cultivated during monsoon season. After hatching, larvae enter the stem by scraping followed by boring through whorl. Following strategies can be followed for prevention and protection of crop:

- Summer plowing of field.

- Destruction of perennating stages in stubbles, cobs, stalks.

- Cut and bury the severely infested plant parts.

- Spray the crop 2–3 weeks after sowing as soon as borer injury to the leaves is noticed with Coragen 18.5 SC (chlorantraniliprole) @ 75 ml using 150 L water/ha with knap-sack sprayer [7, 8, 11].

Shoot fly: Although it is major pest in Southern India but it may infest the maize crop sown in spring season in Northern India. Mainly it attacks the seedling stage of crop (**Figure 9**) where maggots move down to the basal portion through leaf sheath followed by cutting of growing point resulting in dead hearts. Control measures can be as follow:

- Spring crop should be sown between January 20 and February 15.

- Seed should be treated with gaucho (imidacloprid) 600 FS @ 6 ml/kg seed [7].

Figure 8.
Damage of maize crop by maize stem borer.

Figure 9.
Attack of shoot fly in maize crop.

Hairy caterpillar: This pest becomes a serious concern when attacks in epidemic form. They damage the crop by feeding on leaves and soft stem, from gregarious (during younger stage) to distant migration (grown up stage). Prevention and protection strateges can be as follow:

- Collection and destruction of young larvae by cutting and burying the attacked plant parts.
- Physical destruction of large caterpillars [2, 3].

Mite: The attack of mite is serious in June on the young crop or in September–October when the crop is nearing maturity. The affected leaves turn pale and can be recognized from the presence of dusty webs [7, 8].

4.5.2.2 Recently reported pest infestations

In recent years, non-associated pests (**Figures 10** and **11**) have been reported in different parts of India with the details as below (**Table 9**) [3]:

4.5.2.3 Diseases

Seed rot and seedling blight: Poor germination, unthrifty seedlings and seedling mortality are the symptoms. Use disease free seed [7, 8].

Figure 10.
Attack of army worm in maize crop.

Figure 11.
Attack of pollen eating beetle on maize tassels.

Pest name	Plant part infested	Region
Helicoverpa armigera	Cob	Southern India
Chiloloba acuta	Pollen	Northern India

Table 9.
Recently reported pest infestations in maize.

Figure 12.
Maize crop infested with banded leaf and sheath blight.

Figure 13.
Maydis leaf blight attack in maize crop.

Banded leaf and sheath blight: Water soaked, straw colored necrotic lesions alternating with dark brown bands develop on basal leaf sheaths (**Figure 12**). Lesions enlarge and coalesce with each other. Later, sclerotia develop on diseased sheaths, husk and cobs. In severe cases, developing ears are completely damaged and dry up prematurely with cracking of husk. To manage this disease, spray 250 ml Amistar Top 325 SC (azoxystrobin + difenoconazole) in 200 L of water/ha at disease appearance. If needed, repeat the spray at 15 days interval [1–3].

Maydis leaf blight: Symptoms of the disease involve spindle shaped, brownish lesions on the leaves which can further merge to emerge as irregular patches (**Figure 13**). Late sowing, high humidity (>80%) and temperature of 25 + 2°C

Figure 14.
Maize crop attacked by brown stripe downy mildew.

favors the development of disease. Destroy the infected crop residue in the field. Grow improved varieties. Follow spray schedule as against Brown stripe downy mildew [7, 8].

Bacterial stalk rot: Characterized by water soaked appearance and rotting of stem at basal portion causing loss of green color and gives scortching appearance. Rotting of stem results in emitting of foul odor and breakage at 2nd/3rd basal internodes. Excessive rains and poor drainage favors the disease. The infected plants wilt. Destroy the diseased plant debris, keep the fields well drained and use improved varieties for its control [7, 8].

Brown stripe downy mildew: Presence of long, brown colored, interveinal stripes on leaves (**Figure 14**), which if critically watched, have white cottony fungal growth on the lower side of leaves. Whitish downy fungal growth may be observed on close examination on underside of the stripes. Control measures can be as follow:

- Removal of secondary host, that is, *Digitaria sanguinalis*.
- Proper drainage of the fields.
- Spray mancozeb @ 500 g/ha in 250 L of water after about a fortnight of sowing. Give two more sprays at 10-day intervals. Grow recommended varieties [7, 8, 11].

4.6 Harvesting

For use as grain, cobs should be harvested when grains are at about 20% moisture. Whereas to consume as sweet corn, harvesting should be done when tassel starts turning brown and swelling of cob initiates. In case of baby corn, harvest young cob when the silk is near emergence [6].

4.7 Multiple cropping

System in which >2 crops are cultivated in proper sequence on given piece of land during a year. Efficiency of the system is determined by a number of factors namely, manpower, choice of crop/cultivar, availability of irrigation facilities etc.

technical competence, need based farm activities play a critical role in performance of multiple cropping. Following strategies can be adopted for successful adoption of intensive cropping:

- Nursery raising
- Selecting short duration cultivars
- Minimum/no tillage etc. [7, 8].

4.7.1 Intercropping

Maize crop can be cultivated along with other crops as intercrops for better utilization of resources, enhanced income per unit area and time basis. For instance, intercropping of 1 row of fodder cowpea or maize, groundnut and soybean in *kharif* maize sown in 60 cm × 20 cm crop geometry. Apply nutrients to maize as per recommendation and to intercrops on the basis of area under cultivation. Harvest fodder cowpea and maize at about 55 DAS. Furthermore, maize crop can also be cultivated as intercrop in *kharif* blackgram. In this system, maize may be intercropped at every fifth row in the 30 cm apart rows of mash crop. Soybean can be successfully intercropped with maize by sowing one line of soybean between two lines of maize sown at 60 cm [8].

4.7.2 Maize based cropping systems

Crops like wheat, paddy, potato, sugarcane, chickpea, berseem, barley, oats etc. can be grown successfully after harvest of maize crop. Following are some of the most appropriate maize based cropping systems [2, 8]:

- Cowpea/pearl millet/maize (fodder)
- Spring maize-basmati-wheat
- Maize/rice-wheat
- Maize/rice-potato-wheat
- Maize-potato/*toria*-sunflower
- Maize-potato-onion
- Maize-potato-mentha
- Maize-wheat/celery-pearl millet fodder
- Maize/rice-*gobhi sarson*-summer greengram
- Maize-vegetable pea/potato-spring maize
- Maize-potato-sugarcane-wheat
- Maize-wheat-sugarcane

4.8 Techniques to get higher market price

4.8.1 Maize drying

Maize drying is a vital operation which involves removal of moisture from the cobs/grains. It is carried out because high moisture grain will deteriorate rapidly due to grain respiration and heating, germination of grains, mold (fungal) growth and subsequent incidence of mycotoxins (e.g. aflatoxin) and increase insect multiplication and damage. The optimum moisture content of maize should be 14% or less [14].

4.8.1.1 Types of drying

- **Sun drying:** It is a popular method of drying grain where spread grain is exposed to direct sunlight until the desired grain moisture content is achieved. It is low energy cost.

- **Smoking:** The insect infestation is reduced when hung above the fire as the heat reduces the moisture content and the chemicals in smoke deters insect from laying eggs.

- **Air drying:** The maize cobs are hung along the roof of the house to expose it to air and hence the moisture content is minimized [5].

A portable maize dryer 3 ton capacity has been developed by Punjab Agricultural University, Ludhiana as per international norms and recommended to dry maize grains from a moisture level 25 to 15% in 8–10 hours. This cross-flow dryer has three pass, indirect type diesel fired heating system. A control panel to regulate and display the temperature of heated air, exit air and speed of air blower with variable frequency drive is provided for better operation. The dryer can maintain air temperature 60–75°C with the grain temperature of 45°C for seed and 60°C for commercial purpose. The dryer is capable of drying maize grain @ 1.0–1.5% per hour consuming about 4 L/hour. of diesel initially for 1 hour. A provision of heat recovery from flue gases ensures higher fuel efficiency with reduced diesel consumption to about 2 L/hour, later on. The dryer can be operated both with tractor PTO or electricity. One each of skilled and unskilled labor is required to operate this dryer [8].

5. Conclusion

- Adoption of production techniques namely, selection of cultivars, irrigation techniques, INM. IPM and other technological interventions certainly prove propitious in achieving the potential yield targets.

- Maize crop provides better opportunity to scientific community in exploration of resource conservation technologies like zero tillage, partial root drying irrigation, integrated pest management etc.

- Characteristically, maize crop can fit well in diverse crop rotations and intercropping options, which enhances its preference in intensive agriculture.

DOI: http://dx.doi.org/10.5772/intechopen.88997

A. Appendix. Common nutrient deficiency symptoms in maize

Nitrogen deficiency

Phosphorus deficiency

Potassium deficiency

Zinc deficiency

Iron deficiency

Magnesium deficiency

Author details

Manpreet Jaidka[1*], Shikha Bathla[1] and Ramanjit Kaur[2]

1 Krishi Vigyan Kendra, Langroya, Punjab, India

2 Indian Agricultural Research Institute, New Delhi, India

*Address all correspondence to: mjaidka@pau.edu

References

[1] Shukla GN, Kumar A, Jha A, Singh N, Sharma P, Singh J, et al. FICCI PwC knowledge Report: Maize Vision 2022; 2013

[2] Singh C, Singh P, Singh R. Modern Techniques of Raising Field Crops. 2nd ed. New Delhi: Oxford and IBH Publishing Company; 2017. p. 86

[3] Maize Production Technologies in India. Available from: https://farmer.gov.in

[4] Zaidi PH, Jat ML, Jat HS, Lenka D, Swain D. Package of Practices for Profitable Maize Cultivation- a Field Manual. Hyderabad, India: CIMMYT; 2017

[5] Ragasa C, Dankyi A, Acheampong P, Wiredu AN, Chapo-to A, Asamoah M, et al. Patterns of Adoption of Improved Maize Technologies in Ghana. International Food Plocy Rsearch Institute, Working paper 36; 2013

[6] Maize Cultivation: Advanced Maize Production Technologies. Available from: https://www.indiaagronet.com

[7] Anonymous. Package of Practices for Rabi Crops. Ludhiana: Punjab Agricultural University; 2018

[8] Anonymous. Package of Practices for Kharif Crops. Ludhiana: Punjab Agricultural University; 2019

[9] Package of Practices of Maize. Available from: https://kvk.icar.gov.in

[10] Shanahan JM, Rupert J, Schussler J, Groeteke F, Lamm M, Dowgert JV, et al. Subsurface Drip Irrigation for Corn Production and Research. Available from: https://www.pioneer.com

[11] Improved maize production technologies enhanced the net returns of the hill farmers under collaborative programme of ICAR Manipur centre-IIMR, New Delhi

[12] Yang L, Yang H, Jiu-sheng L, Yan-feng L, Hai-jun Y. Estimation of irrigation requirements for drip-irrigated maize in a sub humid climate. Journal of Integrative Agriculture. 2018;**17**(3):60345-60347

[13] Plessis J. Maize Production. Department of Agriculture and obtainable from Resource Centre Directorate Agricultural Information Services Private Bag X144, Pretoria, 0001 South Africa; 2003

[14] Maize Drying and Harvesting. Ento, leaflet 21. Semtokha, Thimphu: National Plant Protection Centre, Department of Agriculture

Chapter 7

Potential and Advantages of Maize-Legume Intercropping System

Sagar Maitra, Tanmoy Shankar and Pradipta Banerjee

Abstract

Intercropping provides enough scope to include two or more crops simultaneously in same piece of land targeting higher productivity from unit area. Maize, a cereal crop of versatile use, as planted in wide rows offers the opportunity for adoption of intercropping. The intercropping system with maize and legume is beneficial in multifaceted aspects. The success of maize-legume intercropping system largely depends on choice of crops and their maturity, density, and time of planting. Advantage of maize-legume combination of intercropping system is pronounced in the form of higher yield and greater utilization of available resources, benefits in weeds, pests and disease management, fixation of biological nitrogen by legumes and transfer of N to associated maize, insurance against crop failure to small holders, and control of erosion by covering a large extent of ground area. Though maize-legume intercropping system exhibits limitations like less scope of farm mechanization, dependence on more human workforce, and chance of achieving less productivity from maize, the system implies more advantages for small holders in developing countries where human workforce is not a constraint. The chapter has focused on beneficial impacts of maize-legume intercropping system.

Keywords: intercropping, maize, legume, advantages, productivity, sustainability

1. Introduction

The cropping system is growing of crops on an area interacting with resources and time and intercropping system is raising of two or more crops simultaneously in the same piece of land [1, 2]. This a common practice in developing countries and it is mostly practiced by small and marginal famers. In tropical world, intercropping is prominently visible with food grain cultivation, whereas in temperate countries forage based intercropping is very much common [3]. Intercropping is generally practiced on small farms with limited resources and it has been observed to enhance yields with greater stability in a variety of crop combinations. Moreover, intercropping system is known by less use of inputs, namely, fertilizers, plant protection chemical, and thus healthy, safe, and high quality food under ecologically sound production system. On-farm biodiversity is also promoted by diversification of crops in through mixed cropping, intercropping and agroforestry systems resulting in variation of diet and net return, higher level of production stability, proper utilization of limited resources human labour-force under low levels of technological intervention [3] and all these ultimately lead to achieve production sustainability in agriculture.

Maize (*Zea mays* L.), also termed as 'corn' and 'queen of cereals', is the third most important cereal of the world, ranks at third position amongst the cereals after rice and wheat and it is a member of Poaceae family. The very cereal has been a staple food for many people in Mexico, Central and South America and parts of Africa. In Europe and rest of the North America, maize is grown mostly as animal feed. Maize is widely cultivated throughout the world having a production of 1147 million tonnes [4]. In various cropping systems as well as in intercropping maize can be fit due to its wider adaptability in different seasons and agro-climatic conditions. Maize is a widely spaced crop and offers ample scope for adoption of intercropping and combination of maize legume in intercropping benefits the agricultural production system by many ways with enhancement of productivity from unit area [2]. This chapter focuses on different aspects of maize based intercropping system, such as considerations, advantages and limitations.

2. Considerations for choosing intercropping system

The success of intercropping depends on different considerations before and during cultivation as because crops grown in mixture may compete spatially and temporally amongst species for available resources. An efficient intercropping system in terms of economic benefits depends on adaptation of planting geometry and choice of compatible and suitable crops. The features of an intercropping system differ with soil and climatic conditions, economic situation and preferences of the farmers. In cereal-legume intercropping system, choice of crop species, density of planting, planting geometry, time of planting and maturity of crops are the key considerations and the success of the system largely depends on these factors.

2.1 Choice of crops

Choice of crops is important in intercropping, because severe competition in mixed culture may not be beneficial and even harmful if proper plant species are not chosen. In this way competition amongst plants can be minimized and better utilization of available resources can be assured. The combination of cereal and legume is considered an ideal because cereals can utilize a portion of biologically fixed nitrogen by legumes. In maize based intercropping system, groundnut is chosen as intercrop maize in South East Asia and Africa [5]. Maize can provide shade to associated legumes and the legume species should be to some extent tolerant to shade. Legume species like black gram (*Vigna mungo*), cowpea (*Vigna unguiculata*), groundnut (*Arachis hypogea*) and green gram (*Vigna radiata*) have much less effect on maize and these are tolerant to maize shade [6, 7]. Cereal-legume intercropping is very common in the continents of Asia, Africa and South America [8], however, in tropical countries, maize based intercropping is practiced with a preference to cowpea [9]. In Central and South America and parts of East Africa intercropping of maize and bean is widely practiced [10]. Maize and dwarf red gram intercropping combination is known as a suitable option in managing cereal component [11].

2.2 Maturity of crop

Maturity of crop is another important consideration in adoption of intercropping. Generally, crops grown in intercropping should have different peak period of growth, otherwise there will be competition amongst the crop species for available resources. The complementary effects benefit the system and these are reflected into yield advantage when the component species in intercropping have different growing period for major demands on available resources. Therefore crops with different duration maturity are chosen to get complementary effects. Maize has been recognized as a common crop in cereal-based intercropping and treated as base crop in additive series and dissimilar legumes are preferably considered as intercrop. In maize-based intercropping system choosing short duration legumes as intercrops is an ideal option. For example, in maize + green gram intercropping system, initial growth of maize is slow and it reaches at knee-height stage after 6–7 weeks and peak light demand starts from 55 to 60 days after sowing and by this period green gram sown at the same time will be in reproductive stage or in close to harvest. In this way green gram completes its major growth period and maize starts the same and thus high level of complementarity is observed.

2.3 Plant density and maturity of component crops

Optimum plant stand is synonymous to optimum yield. But in intercropping system two or more crops are accommodated in the same land at the same time and thus there may be reduction in population of crops compared to pure stand of individual species. On the basis of plant density, intercropping may be categorized into two groups, namely, additive series and replacement series. The additive series is comprised of addition of intercrop within fullest population of base crop. Another crop known as intercrop and it is sown into the base crop population by adjusting row spacing or changing planting geometry. Sometimes, paired row planting of maize is done to accommodate greater space for intercrops. But in replacement series of intercropping, there is not the concept of base crop and the crops (two or more) considered are termed as component crops or intercrops. In such type of intercropping, introduction of a component crop is made by replacing another and none of the component crops are sown with 100% population as recommended in their pure stands. It is very clear that certain proportion of population of one crop component is sacrificed and another component is introduced in that place. In many intercropping situations with replacement series, yield advantages are maximized by increasing population density in excess than their recommended population in the sole cropping. Here, the competition is relatively lesser in between component crops as compared to additive series. As maize is widely spaced crop and generally row spacing ranges between 60 to 90 cm and intercrops can easily be raised in uniform rows of planting. The planting geometry, particularly, paired row planting of maize may enhance the efficiency of growth parameters as well as yield of maize and associated legumes by efficient accommodation of crops. Prasad and Brook [12] observed an enhanced LAI per unit area with increase in plant population of maize in maize-soybean intercropping system. Under the major demand for resources at different times of system duration, the long duration cereal crop maize can recover its resource needs in combination with short duration legumes during remaining phase of growth that is after harvest of legumes [13].

2.4 Time of planting

Maize is recognized as a very common crop in intercropping system in which legumes can be sown easily. Generally, in maize based intercropping systems, as maize has slower initial growth rate up to knee height stage (6–7 weeks of sowing), if short duration legumes are sown simultaneously can reach into their reproductive stage can start their reproductive period and hence competition for common natural resources do not appear at the same period. Maize has diverse use and if maize is considered as fodder in intercropping, competition does not come into figure because of enhance biomass yield and mixture of grass-legume combination enhances the quality of forage in terms of dietary value. Moreover, maize has higher potential for accumulation of carbohydrate, a source of energy as fodder, from unit area on daily basis. However, legumes can be planted in maize at the same time can also register higher growth attributes because of wider spacing of maize as grain crop.

3. Advantages of intercropping

Maize and legume intercropping system has advantages in many ways and so preferred by small and marginal farmers. Experimental results showed that maize-legume intercropping can assure higher yield, soil restoration and greater impact of system productivity.

3.1 Advantage in improving productivity and soil fertility

In Intercropping, more crops are grown simultaneously in unit area which results not only greater productivity but also utilizes natural resources more efficiently. Management of pests, diseases and weeds is easier because of less incidence which leads to greater yield. Another important aspect of maize-legume intercropping is restoration of soil fertility.

3.1.1 Higher yield and greater resource utilization

Yield is the basic consideration for assessing benefits of intercropping. In maize-legume intercropping maize is treated as based crop without much variation in plant stand of cereal component. In additive series of intercropping, legumes add plant population per unit area and benefits are achieved as total yield of crops, namely maize and legume yields. Further, in a combination of legume and non-legume, generally non-legume component is benefited by sharing atmospheric nitrogen fixed by legumes. In assessing efficiency of an intercropping system, some competition functions are considered. Of which land equivalent ratio (LER) is a very common index used to measure productivity of intercropping system. Willey and Osiru [14] proposed the concept of the LER and it is defined as the proportionate land area required under pure stand of crop to produce the same productivity as obtained in an intercropping at the same management level. Actually, LER is the summation of ratios of the yield of each crop species involved in intercropping system to its corresponding pure stand yield. Experiments carried out in different countries clearly exhibited higher LER values in maize-legume intercropping system (**Table 1**).

The LER indicates the advantage of an intercropping with efficient resource utilization compared to pure stands of respective crops. The value of LER greater than unity (1.0) is indicative of the advantages in intercropping system [2].

The LER indicates on efficiency of using land area, but time factor is not considered for which the crop occupies the land area. To rectify the limitation of the LER, the concept of area time equivalent ratio (ATER) has been developed considering the occupancy of land by the crops for certain periods [23]. Like the LER, values of the ATER more than unity also indicate advantage of intercropping. Different researchers noted beneficial ATER values with maize-legume intercropping systems (**Table 2**).

However, researchers comment that the LER overestimates and the ATER underestimates the land-use efficiency [27].

Crop equivalent yield is another expression for evaluating the efficiency of intercropping system [25]. Actually, in maize-legume intercropping system, total

Intercropping system	Ratio/proportion	LER	Country	References
Maize + bean	2:1	2.60	Kenya	[15]
Maize + cowpea	1:1	1.72	Turkey	[16]
Maize + French bean	1:2	1.66	India	[17]
Maize + soybean	1:1	1.54	Nigeria	[18]
Maize + groundnut	2:2	1.42	Ghana	[19]
Maize + garden pea	1:2	1.56	Bangladesh	[20]
Maize + soybean	100% + 75%	1.60	Turkey	[21]
Maize + groundnut	2:2	1.82	India	[7]
Maize + soybean	2:2	1.90	China	[22]

Table 1.
Land equivalent ratio (LER) in maize-legume intercropping systems.

Intercropping system	Ratio	ATER	Country	References
Maize + black gram	1:2	1.37	India	[24]
Maize + black gram	1:2	1.47	India	[17]
Maize + soybean	2:6	1.32	India	[25]
Maize + black cowpea	2:2	1.51	India	[26]

Table 2.
Area time equivalent ratio (ATER) in maize-legume intercropping systems.

yields are converted in the form of base crop (maize) equivalent yield by considering the intercrop yield and market price of maize (base crop) and associated intercrops. In maize-legume intercropping system it is termed as maize equivalent yield (MEY) and expressed in kg^{-ha}. If the base crop equivalent yield is obtained higher in intercropping combinations than base crop yield, then intercropping is considered advantageous. **Table 3** indicates advantageous MEYs as obtained by the researchers in experiments.

The greater yields in intercropping is recorded when the component crops show complementary effects amongst themselves and use natural resources efficiently than raised as sole crops [28]. The crops with inherent capability can only utilize natural resources efficiently and complementarity plays important role in resource utilization [2]. Further, higher yield of both the crops in maize-cowpea intercropping combination was noted than pure stands [29].

In soils with low nitrogen content, maize legume intercropping performed well [30]. Yield advantage in intercropping is expressed by crops because of greater use of growth resources like light, water, and nutrients and this efficient use is

Intercropping system	Ratio	MEY (kg^{-ha})	Sole maize yield (kg^{-ha})	Sole legume yield (kg^{-ha})	References
Maize + soybean	2:6	9470	7092	5450	[25]
Maize + black cowpea	2:2	7699	5062	4785	[26]
Maize + garden pea	1:2	20,220	8200	6450	[20]

Table 3.
Maize-equivalent yield (MEY) in maize-legume intercropping systems.

converted into biomass [2, 31]. The combination of maize-cowpea intercropping can assure greater light interception and check evaporation loss of soil moisture than pure stand of maize [32].

Maize and legumes are morphologically dissimilar and their time of peak demand and requirement of light, nutrients and water are different. Therefore, complementary effect between component crops is very common. Jiao et al. [33] noted that maize used strong light and groundnut preferred weak light (because maize provided partial shade) in maize-groundnut intercropping system and the system registered yield advantage. Further, soybean-maize intercropping has been known for efficient utilization of light, nutrients and available soil moisture [2, 34]. Soil moisture or water availability to plants is a determining factor in intercropping systems and efficient water use leads to use of other resources. Cereal-legume combination is known to use available water resources more efficiently than pure stands of crops. Scientific investigations showed that maize-legume combination registered greater water use efficiency than that of sole crops and under water stress conditions, it could be one of the best options [35] as soybean as a deep rooted crop having efficiency to use soil moisture from deeper layer (below 1 m) of the soil [36].

3.1.2 Weed management

Intercropping is an effective practice for weed management because enough of ground area is covered by crops which suppress weed growth. Combination of maize and legumes in intercropping is known to reduce weed population and weed biomass compared to pure stands of maize. Research evidences clearly show benefits of intercropping as it provides competitive effect against weeds both spatially and temporally than pure stands of maize. Reduced weed growth in maize-cowpea intercropping system than sole cropping of maize. Chalka and Nepalia [37] mentioned that in maize-legume intercropping systems, maize + cowpea and maize + soybean reduced NPK removal through weeds by 37.4 and 38.0% respectively and the two intercropping combinations registered higher biological yield of maize. Rahimi et al. [38] reported that maize-black gram intercropping combination of either 1:1 or 2:2 recorded lower densities of total weeds compared to pure stand of maize. Shah et al. [39] opined that weed smothering efficiency was higher in intercropping of maize with soybean than the combination of maize with green gram and it was due to the lower availability of space and light leads to reduce the weeds population with maize-soybean intercropping system. Weed biomass is reduced in intercropping as reported by researchers for maize–legume combinations [40, 41]. In studies it has been claimed that enhancement of diversity of crop species in intercropping system maintains a highly asymmetric competition over weeds resulting in less weed biomass [42, 43]. Weeds compete with crops for available resources and less weed occurrence assures ultimately higher productivity.

3.1.3 Pest and disease management

Intercropping systems can influence the pest and pathogen population dynamics. The population of beneficial insects such as parasites and predators are enhanced in polyculture due to diversity of crops [2] and presence of harmful pests may remain below the economic threshold level. Thus, plant protection becomes easy and use of chemicals for crop protection comes down which ultimately reduces the chemical pollution to agricultural ecology, however, monoculture of maize requires more chemical pesticides [44]. In intercropping system, two or more crop species are cultivated which creates complexity in food and habitat of pests. Further, intercropping of maize with legumes is known to increase population of beneficial insects and decrease the population of bud worm, corn borer, leaf hopper and maize stalk borer [1, 45]. The intercropping system has also an impact against disease management, because in mixture of crops functional diversity is created that checks population increase of pathogen. Some diseases of legume crops like angular leaf spot (*Phaeoisariopsis griseola*) of beans and ascochyta blight (*Mycosphaerella pinodes*) were observed with less severity when these were intercropped with maize [46, 47] than pure stand of legumes. Reduction of pest-disease incidence not only saves the crop loss and better yield but also assures less use of chemicals for plant protection and thus minimizes the chance of pollution in crop ecology.

3.1.4 N-fixation by legumes and transfer to associated non-legume

Legumes are known to fix atmospheric nitrogen biologically. The biological nitrogen fixation (BNF) is a process where some bacteria convert atmospheric N_2 into ammonia (NH_3) and making it available to plants. In maize-legume intercropping system, both the crops acquire N from the common soil pool and compete and thus deficit of mineral N may occur in the rhizosphere which promotes legume to fix atmospheric N [48, 49]. Maize is an exhaustive crop and legumes are soil replenishing crops and decomposition of legumes residue improves soil fertility. In the soils with poor available nitrogen status, the biologically fixed nitrogen plays an important role. Under the situation of limited supply of nitrogenous fertilizer also intercropping legume and non-legume may a suitable option of nutrient management. Further, chemical N fertilizers are responsible for degradation of ecosystem in the form of nitrate pollution and legumes grown as intercrops help in environmental sustainability [50]. In maize-soybean intercropping system, soybean supplements nitrogen to cereal component [51]. Maize grown as forage in intercropping with legumes is known to improve quality parameters of forage like higher crude protein and mineral content and digestibility [48, 52]. Biologically fixed N by pigeon pea was transferred to associated maize and N content and uptake by maize was improved in maize-pigeon pea intercropping system [53]. The associated non-legume crop (maize) gets benefit of fixed N by legumes [1]. Thus, maize-legume intercropping system is beneficial in terms of N economy too. Leaf defoliation of legumes is known to increase productivity of maize–soybean intercropping system [22].

3.1.5 Erosion control

Intercropping is advantageous in terms of erosion control because of coverage of more ground area than monocropping of cereals. The striking actions of rain drops can erode the bare or uncovered soil, but the coverage of soil by legumes can check it. In maize-cowpea intercropping combination, ground area is mostly covered, thus soil erosion is reduced [54]. Taller crop like maize also plays a vital role as wind break and protects the crops with shorter canopy (like legumes) as well as erosion caused by wind [45].

3.2 Advantage in enhancement of system productivity

3.2.1 Insurance against crop failure to small holders

Intercropping is a common practice of small and marginal farmers in developing countries of Asia and Africa and in risky and fragile ecological conditions which is known as a suitable practice to provide natural insurance and thus provides a profitable shape to farm economy. Under moisture stress conditions, more of ground area is covered under maize-legume intercropping than sole cropping of maize which leads to less evaporation loss of soil moisture. Under extreme conditions, may be due to either biotic or abiotic factors, a crop may fail, but there will be less chance of failure of more crops grown in intercropping, which are morphologically dissimilar and if so happened some yield and return will be earned to save small holders' economic interest. Thus stability in yield and return are achieved due to creation of crop diversity in the intercropping systems. In economic point of view, it may be stated that small farmers may face problem of seasonal price variability of commodities which often can destabilize net realization, but diversification in the form of intercropping can stabilize farm income to a great extent. Experimental results indicated superiority of intercropping maize-beans in soil fertility restoration and income enhancement than monocropping of the component crops [55]. Yield enhancement of crops is another basis to strengthen the economy of small and marginal farmers adopting intercropping system [56]. Though intercropping of maize-grain legumes is labour and cost intensive, small holders of central Mozambique prefer it because of reduced risk of crop failure and enhanced productivity [57].

3.2.2 Sustainability of the system

Intercropping is now in the centre of attention targeting sustainability in agriculture. The negative impacts of industrialized and modern agriculture have already been realized and issues are very crucial in crop ecology to achieve sustainability. On the other side, maize-legume intercropping has enough potential in the form of more yields from limited resource, proper utilization of resources, and restoration of soil fertility, efficient pest management and creation of above and below ground diversity. In the moisture stress or resource poor conditions, intercropping provides natural insurance against crop failure caused by biotic and abiotic factors and thus ascertains economic stability of small holders. Considering the multiple advantages, it can be stated that maize-legume intercropping system is one of the suitable options for achieving production sustainability for small holders.

4. Limitations of intercropping

Despite a number of benefits of maize-legume intercropping over monocropping, sometimes intercropping may exhibit some limitations especially in terms of agronomic management. In the field where farm mechanizations have been adopted, intercultural operations and harvest become difficult with two dissimilar crops. However, there is no problem where the intercrops are harvested for forage or grazed [13]. It may be mentioned that where human workforce is sufficient, particularly in developing countries, there is no need for investment in costly machines for agronomic management and harvest of crops in intercropping and in this regard intercropping does not express any disadvantages. Intercropping may cause yield loss of the base crop (maize) compared to its sole stand, but MEY become more and thus intercropping may be considered more productive than monoculture. Further in intercropping, crowded crop canopy may create a microclimate which may be congenial to spread of fungal pathogens, but in maize-legumes intercropping combination, such incidences are not common.

5. Conclusion

Considering the importance of maize in cereal basket of the world, production sustainability is a prime concern. Maize is an exhaustive crop by nature that requires enough nutrient inputs to achieve target yield. Under small holders' practices in poor soil and fragile ecological conditions continuous growing of maize may create further depletion of soil nutrients causing a threat to production sustainability. In this regard, maize-legume intercropping system is considered a suitable option as it has enough potential to replenish the soil nutrients, produce more yield and economic benefit by utilizing limited resource, check damage caused by pests, diseases and weeds to a large extent, control soil erosion by covering ground and provide natural insurance to small holders under risky conditions against crop failure. Thus, in true sense, maize-legume intercropping system can boost yield as well as production sustainability of the system as a whole.

Author details

Sagar Maitra*, Tanmoy Shankar and Pradipta Banerjee
M.S. Swaminathan School of Agriculture, Centurion University of Technology and Management, Odisha, India

*Address all correspondence to: sagar.maitra@cutm.ac.in

References

[1] Seran TH, Brintha I. Review on maize based intercropping. Journal of Agronomy. 2010;**9**(3):135-145. DOI: 10.3923/ja.2010.135.145

[2] Maitra S, Palai JB, Manasa P, Kumar DP. Potential of intercropping system in sustaining crop productivity. International Journal of Agriculture, Environment and Biotechnology. 2019;**12**(1):39-45. DOI: 10.30954/0974-1712.03.2019.7

[3] Anil L, Park J, Phipps RH, Miller FA. Temperate intercropping of cereals for forage: A review of the potential for growth and utilization with particular reference to the UK. Grass and Forage Science. 1998;**53**:301-317. DOI: 10.1046/j.1365-2494.1998.00144.x

[4] FAOSTAT. Maize Production. Crops/Region/Production Quantity from Pick Lists, United Nations, Statistics Division [Internet]. 2018. Available from: http://www.fao.org/faostat/en/#data/QC [Accessed: 10 February 2020]

[5] Shalim Uddin M, Rahaman MJ, Bagum SA, Uddin MJ, Rahaman MM. Performance of intercropping of maize with groundnut in saline area under rainfed condition. Pakistan Journal of Biological Sciences. 2003;**6**(2):92-94. DOI: 10.3923/pjbs.2003.92.94

[6] Mandal MK, Banerjee M, Banerjee H, Alipatra A, Malik GC. Productivity of maize (Zea mays) based intercropping system during kharif season under red and lateritic tract of West Bengal. The Bioscan. 2014;**9**(1):31-35

[7] Manasa P, Maitra S, Reddy MD. Effect of summer maize-legume intercropping system on growth, productivity and competitive ability of crops. International Journal of Management, Technology and Engineering. 2018;**8**(12):2871-2875

[8] Layek J, Das A, Mitran T, Nath C, Meena RS, Yadav GS, et al. Cereal+legume intercropping: An option for improving productivity and sustaining soil health. In: Meena RS et al., editors. Legumes for Soil Health and Sustainable Management. Singapore: Springer Nature; 2018. pp. 347-386. DOI: 10.1007/978-981-13-0253-4_11

[9] Van Kessel C, Roskoski JP. Row spacing effects on nitrogen fixation, nitrogen yield and soil nitrogen uptake of intercropped cowpea and maize. Plant and Soil. 1988;**111**:17-23

[10] Singh SP, Singh RA, Singh UN. Yield and economics of sesame and blackgram, intercropping under dryland conditions of Vindhyan plateau, India. Indian Journal of Agronomy. 1998;**43**:321-324

[11] Gallo ADS, Fontanétti A, Guimarães NDF, Morinigo KPG, de Souza MDB, da Silva RF, et al. Agronomic characteristics, nutritional status and yield of corn intercropped with dwarf pigeon pea in different spatial arrangements of plants. Revista de Ciências Agrárias. 2018;**41**(2):356-366. DOI: 10.19084/RCA17297

[12] Prasad RB, Brook RM. Effect of varying maize densities on intercropped maize and soybean in Nepal. Experimental Agriculture. 2005;**41**:365-382. DOI: 10.1017/S0014479705002693

[13] Lithourgidis AS, Dordas CA, Damalas CA, Vlachostergios DN. Annual intercrops: An alternative pathway for sustainable agriculture. Australian Journal of Crop Science. 2011;**5**(4):396-410

[14] Willey RW, Osiru DSO. Studies on mixtures of maize and beans (*Phaseolus vulgaris*) with particular reference

to plant population. The Journal of Agricultural Science. 1972;**79**:517-529

[15] Odhiambo GD, Ariga ES. Effect of intercropping maize and beans on striga incidence and grain yield. In: Proc. 7th Eastern Southern Africa Regional Maize Conf., 11-15 February 2011. Vol. 7. 2011. pp. 183-186

[16] Yilmaz S, Atak M, Erayman M. Identification of advantages of maize-legume intercropping over solitary cropping through competition indices in the East Mediterranean region. Turkish Journal of Agriculture. 2008;**32**:111-119

[17] Choudhary VK. Suitability of maize-legume intercrops with optimum row ratio in mid hills of eastern Himalaya, India. SAARC Journal of Agriculture. 2014;**12**(2):52-62. DOI: 10.3329/sja.v12i2.21916

[18] Kamara AY, Tofa AI, Ademulegun T, Solomon R, Shehu H, Kamai N, et al. Maize-soybean intercropping for sustainable intensification of cereal-legume cropping system in northern Nigeria. Experimental Agriculture. 2017;**55**(1):73-87. DOI: 10.1017/S0014479717000564

[19] Kermah M, Franke AC, Samuel AN, Benjamin DKA, Abaidoo RC, Giller KE. Maize-grain legume intercropping for enhanced resource use efficiency and crop productivity in the Guinea savanna of northern Ghana. Field Crops Research. 2017;**213**:38-55. DOI: 10.1016/j.fcr.2017.07.008

[20] Khan MAH, Sultana N, Akter N, Zaman MS, Islam MR. Intercropping gardenpea (Pisium sativum) with maize (Zea mays) at farmers field Bangladesh. Journal of Agricultural Research. 2018;**43**(4):691-702. DOI: 10.3329/bjar.v43i4.39166

[21] Aydemir SK. Maize and soybean intercropping under different seed rates of soybean under ecological condition of Bilecik, Turkey. International journal of Environmental Science and Technology. 2018. DOI: 10.1007/s13762-018-1986-2

[22] Raza MA, Khalid MHB, Zhang X, Feng LY, Khan I, Hassan MJ, et al. Effect of planting patterns on yield, nutrient accumulation and distribution in maize and soybean under relay intercropping systems. Scientific Reports. 2019;**9**:4947. DOI: 10.1038/s41598-019-41364-1

[23] Hiebsch CK. Interpretation of yields obtained in crop mixture. In: Agronomical Abstracts, American Society of Agronomy. Madison, Wisconsin; 1978. p. 41

[24] Kheroar S, Patra BC. Advantages of maize-legume intercropping systems. Journal of Agricultural Science and Technology. 2013;**3**:733-744

[25] Yogesh S, Halikatti SI, Hiremath SM, Potdar MP, Harlapur SI, Venkatesh H. Light use efficiency, productivity and profitability of maize and soybean intercropping as influenced by planting geometry and row proportion. Karnataka Journal of Agricultural Sciences. 2014;**27**(1):1-4. DOI: 10.23910/IJBSM/2017.8.6.3C0363

[26] Jan R, Saxena A, Jan R, Khanday M, Jan R. Intercropping indices and yield attributes of maize and black cowpea under various planting pattern. The Bioscan. 2016;**11**(2):1-5. (Supplement on Agronomy)

[27] Maitra S, Ghosh DC, Sounda G, Jana PK, Roy DK. Productivity, competition and economics of intercropping legumes in finger millet (*Eleusine coracana*) at different fertility levels. Indian Journal of Agricultural Sciences. 2000;**70**(12):824-828

[28] Willey RW. Intercropping—Its importance and research needs. Part 1: Competition and yield advantages. Field Crops Research. 1979;**32**:1-10

[29] Kimou SH, Coulibaly LF, Koffi BY, Toure Y, Dede KJ, Kone M. Effect of row spatial arrangements on agromorphological responses of maize (*Zea mays* L.) and cowpea (*Vigna unguiculata* L.). African Journal of Agricultural Research. 2017;**12**(34):2633-2641. DOI: 10.5897/AJAR2017.12509

[30] Vesterager JM, Nielsen NE, Hogh-Jensen H. Effects of cropping history and phosphorus source on yield and nitrogen fixation in sole and intercropped cowpea-maize systems. Nutrient Cycling in Agroecosystems. 2008;**80**:61-73. DOI: 10.1007/s10705-007-9121-7

[31] Reddy AS, Palled YB. Effect of intercropped fodder cowpea on maize and system productivity in maize + fodder cowpea intercropping systems. Journal of Farm Science. 2016;**29**(2):265-267

[32] Ghanbari A, Dahmardeh M, Siahsar BA, Ramroudi M. Effect of maize (*Zea mays* L.)—cowpea (*Vigna unguiculata* L.) intercropping on light distribution, soil temperature and soil moisture in and environment. Journal of Food, Agriculture and Environment. 2010;**8**:102-108. DOI: 10.1234/4.2010.1458

[33] Jiao NY, Zhao C, Ning TY, Hou LT, Fu GZ, Li ZJ, et al. Effects of maize-peanut intercropping on economic yield and light response of photosynthesis. Chinese Journal of Applied Ecology. 2008;**19**:981-985

[34] Raji JA. Intercropping soybean and maize in a derived Savana ecology. African Journal of Biotechnology. 2007;**6**(16):1885-1887

[35] Tsubo MS, Walker, Ogindo HO. A simulation model of cereal-legume intercropping systems for semi-arid regions: II. Model application. Field Crops Research. 2005;**93**:23-33. DOI: 10.1016/j.fcr.2004.09.002

[36] Maitra S. Soyabean. In: Hedayetullah M, Zaman P, editors. Forage Crop of the World. Vol. **2**. USA: Apple Academic Press; 2018. pp. 101-120

[37] Chalka MK, Nepalia V. Nutrient uptake appraisal of maize intercropped with legumes and associated weeds under the influence of weed control. Indian Journal of Agricultural Research. 2006;**40**:86-91

[38] Rahimi IM, Amanullah M, Ananthi T, Mariappan G. Influence of intercropping and weed management practices on weed parameters and yield of maize. International Journal of Current Microbiology and Applied Sciences. 2019;**8**(4):2167-2172. DOI: 10.20546/ijcmas.2017.604.175

[39] Shah SN, Shroff JC, Patel RH, Usadadiya VP. Influence of intercropping and weed management practices on weed and yields of maize. International Journal of Science and Nature. 2011;**2**(1):47-50

[40] Bilalis D, Papastylianou P, Konstantas A, Patsiali S, Karkanis A, Efthimiadou A. Weed-suppressive effects of maize-legume intercropping in organic farming. The International Journal of Pest Management. 2010;**56**:173-181. DOI: 10.1080/09670870903304471

[41] Stoltz E, Nadeau E. Effects of intercropping on yield, weed incidence, forage quality and soil residual N in organically grown forage maize (*Zea mays* L.) and faba bean (*Vicia faba* L.). Field Crops Research. 2014;**169**:21-29. DOI: 10.1016/j.fcr.2014.09.004

[42] Connolly J, Sebastia MT, Kirwan L, Finn JA, Llurba R, Suter M, et al. Weed suppression greatly increased by

plant diversity in intensively managed grasslands: A continental-scale experiment. Journal of Applied Ecology. 2018;**55**:852-862. DOI: 10.1111/1365-2664.12991

[43] Elsalahy H, Döring T, Bellingrath-Kimura S, Arends D. Weed suppression in only-legume cover crop mixtures. Agronomy. 2019;**9**:648. DOI: 10.3390/agronomy9100648

[44] Singh BB, Adjeigbe HA. Improving cowpea cereals-based cropping systems in the dry Savannas of West Africa. In: Fatokun A, Tarawali SA, Singh BB, Kormawa PM, Tamo M, editors. Challenges and Opportunities for Enhancing Sustainable Cowpea Production. Ibadan, Nigeria: IITA; 2002. pp. 276-284

[45] Kinama MJ, Pierre HMJ. A review on advantage of cereals-legumes intercropping system: Case of promiscuous soybeans varieties and maize. International Journal of Agronomy and Agricultural Research. 2018;**12**(6):155-165

[46] Vieira RF, de Paula TJ Jr, Teixeira H, Vieira C. Intensity of angular leaf spot and anthracnose on pods of common beans cultivated in three cropping systems. Ciénc. Agrotec. 2009;**33**:1931-1934. DOI: 10.1590/S1413-70542009000700040

[47] Schoeny A, Jumel S, Rouault F, Lemarchand E, Tivoli B. Effect and underlying mechanisms of pea-cereal intercropping on the epidemic development of ascochyta blight. European Journal of Plant Pathology. 2010;**126**:317-331. DOI: 10.1007/s10658-009-9548-6

[48] Dwivedi A, Dev I, Kumar V, Yadav RS, Yadav M, Gupta D, et al. Potential role of maize-legume intercropping systems to improve soil fertility status under smallholder farming systems for sustainable agriculture in India. International Journal of Life Sciences Biotechnology and Pharma Research. 2015;**4**(3):145-157

[49] Massawe PI, Mtei KM, Munishi LK, Ndakidemi. Existing practices for soil fertility management through cereals-legumes intercropping systems. World Research Journal of Agricultural Sciences. 2016;**3**(2):80-91

[50] Fustec J, Lesuffleur F, Mahieu S, Cliquet JB. Nitrogen rhizo deposition of legumes—A review. Agronomy for Sustainable Development. 2010;**30**:57-66. DOI: 10.1051/agro/2009003

[51] Owusu A, Sadick A. Assessment of soil nutrients under maize intercropping system involving soybean. International Research Journal of Agricultural and Food Sciences. 2016;**1**(3):33-43

[52] Dahmardeh M, Ghanbari A, Syasar B, Ramrodi M. Intercropping maize (zea mays L.) and cow pea (*Vigna unguiculata* L.) as a whole-crop forage: Effects of planting ratio and harvest time on forage yield and quality. Journal of Food, Agriculture and Environment. 2000;**7**:505-509. DOI: 10.3923/AJPS.2009.235.239

[53] Olujobi OJ, Oyun MB, Oke DO. Nitrogen accumulation, growth and yield of maize in pigeon pea/maize intercrop. Global Journal of Biology, Agriculture and Health Sciences. 2013;**2**(1):42-48

[54] Kariag BM. Intercropping maize with cowpeas and beans for soil and water management in Western Kenya. In: Proc. of the 13th International Soil Conservation Organization Conference, July 4-9, 2004. Brisbane: Conserving Soil and Water for Society; 2004. pp. 1-5. DOI: 2010.135.145

[55] Onduru DD, Du Preez CC. Ecological and agro-economic study

of small farms in sub-Saharan Africa. Agronomy for Sustainable Development. 2007;**27**:197-208. DOI: 10.1051/agro:2007003

[56] Lulie B. Intercropping practice as an alternative pathway for sustainable agriculture: A review. Academic Research Journal of Agricultural Science and Research. 2017;**5**(6):440-452. DOI: 10.14662/ARJASR2017.057

[57] Rusinamhodzia L, Corbeels M, Nyamangara J, Giller KE. Maize–grain legume intercropping is an attractive option for ecological intensification that reduces climatic risk for smallholder farmers in Central Mozambique. Field Crops Research. 2012;**136**:12-22. DOI: 10.1016/j.fcr.2012.07.014

Chapter 8

Chemical Weed Management in Maize (*Zea mays* L.) under Conservation Agricultural Systems: An Outlook of the Eastern Gangetic Plains in South-Asia

Akbar Hossain, Mst. Tanjina Islam, Md. Shohidul Islam, Nurislam, Sharif Ahmed, Khokan Kumer Sarker and Mahesh Kumar Gathala

Abstract

Maize is a widely grown cereal after rice and wheat and contributes almost 5% to the global dietary supply. In the Eastern Gangetic Plains (EGP) including India, Bangladesh, and Nepal, maize is an emerging cash crop, because of its high yield potentiality and also the favorable climatic conditions which allow maize production round the year. In Bangladesh, area and production of maize are escalating due to the increasing demand for poultry, livestock, and fish feed, and fodder for animals and starch industries in the region. Presently, more than 90% of maize is planted by manual dibbling following 5–6 intensive tillage, which increases the cost of cultivation. The conservation agricultural (CA)-based new agricultural practices could overcome those above challenges. CA is cost-effective and environmentally friendly; however, weeds are one of the key challenges in the system. The chapter described the uses of herbicides in different ways of combinations to make effective weed control in CA-based maize to achieve potential production and profits by reducing the intensive pressure of manual weeding. The efficient and right use of pre-plant/sowing, pre- and post-emergence herbicides and their combination may be the best way for effective control of weeds in maize production.

Keywords: conservation agriculture, weed management, herbicides, maize

1. Introduction

Maize (*Zea mays* L.) is the third most important cereal after rice and wheat, which is widely grown in the world and used as a primary staple food in many developing countries. The area and production under maize in the world in 2013–2014 was 177 m ha with 967 Mt. production and contributed almost 5% of the world's dietary energy supply [1]. Recent projection indicates that by 2020 the demand of maize in all developing countries will overtake the demand of wheat and rice [2], with Asia accounting for nearly 60% of the global demand for maize. Half of the world maize is produced in the developing countries where maize grain is one of the major sources of food energy for the poor people and its plant biomass provides feed for animals [3].

Eastern Gangetic Plains (EGP) covering three neighboring countries (India, Bangladesh and Nepal) has a very high yield potential of maize production. In recent decades, maize an emerging as a cash crop for smallholder farmers because of its high yield potentiality due to favorable climatic conditions which allows maize production round the years in three seasons [4]. The EGP is mainly dominated by rice-based farming systems having maximum coverage of rice-rice cropping system which is associated with own problems of high water consumption, production costs and labour use, and further soil health deterioration. All these associated problems make rice-rice production is unsustainable and unprofitable. Farmers are looking for the alternative crop of dry season rice called boro rice (winter rice) to diversify their cropping systems and maize is one of the most suitable crops to provide high yield and profit margin in a more sustainable way.

Presently, more than 90% maize grows in EGP by manual dibbling after intensive tillage operations (4–6 dry tillage) which delayed the maize sowing by at least 1–2 weeks [5]. A number of repetitive tillage operations increase the cost of cultivation due to higher energy and labor use and delayed planting which further reduces the net margins of the farmers. Since, the conservation agricultural (CA) based management practices (minimum/zero tillage) are the new emerging technologies growing an attention worldwide due to higher economic benefits, improvement in soil health and also found more environmental friendly by reducing greenhouse gases.

Although CA-based maize cultivation has many benefits, however weed infestation is one of the serious constraints that limits maize production though weed management is also a major concern in conventional production system. The competition between weeds and maize at critical growth stages could be reduced both the quality and quantity of maize yield over 30% [6] as weeds compete with crop for essential resources [7–9]. For controlling weeds in crop field, farmers are generally adopting mechanical, cultural, biological and chemical control methods. Among them, exhausted by cultural methods, farmers are moving towards other alternative methods due to labour crisis during critical period of weed control [6, 10, 11]. While, the mechanical methods are still useful but are unable to effective control of weeds successfully due to the absence of right machinery [12]. The judicious and right use of a different combination of herbicides as pre-plant, pre-emergence and post-emergence can provide effective and efficient weed control under CA system as well as conventional system [13]. The chemical weed control may be provided cost-effective, faster and better weed control [14–16].

2. Importance of tillage options on weed infestation in the crop field

Weed management strategies are the key to success of the conservation agriculture based crop production especially at the beginning of 2 years. With the shift from conventional agriculture to conservation agriculture there will be a shift dynamic change in weed species, population and diversity. Therefore, initial weed management is an important phenomenon through integrated weed management approaches like stale seed-bed, mulch, and adjustment in planting dates, biological and mechanical control and effective combination use of herbicides. The many long-term studies clearly indicated that if the proper weed management options adopted at initial 2 years then the weed population significantly reduced with time under CA-based practices than conventional practices [17–19]. Tillage systems affect the composition of weeds in a field [20]. Typically, perennial weeds in conventional tillage systems cannot be fixed due to repeated plowing but alteration of soil between the surface to sub-surface exposes new weed seeds. In the other hand, the perennial weeds are more prominent initially in CA-based tillage but also broadleaf weeds may be prominent due to the light-weight and density of seeds which remain on the surface facilitate for conducive germination environment. The several studies showed that *Phalaris minor* populations significantly reduced under zero tillage but broadleaves especially *Rumex* found increases [21–23]. A recent study in the EGP suggested, if initial weeds are not managed properly then *Polygonum* species and sedges dominated under CA systems. However, perennial weeds in these systems, the plant roots are not pulled out shall be subject to the complete eradication [24] may regenerate again. [25] reported that dry weight of weeds in crops without plowing method compared to conventional tillage crops decreased by respectively 61 and 77%. Another earlier findings [26] observed that nutrients uptake by crops under CA systems was enriched when weeds was in CA-system were reduced. CA-based agricultural systems conserve the soil health through improving the soil organic matter as well as soil microbial finally leading for the higher productivity of crops [27, 28].

3. Major weed species in association with maize

Infesting weed species in crop field generally depend on crop species, cropping systems and their management practices, environmental conditions, seasonal variation, soil properties, nutrients and moisture status as well as soil types. The Eastern Gangetic Plain (EGP) as known for high weed seed bank in soils especially lower EGP (Bangladesh and West Bengal) due to flash and steady floods bring weed seeds from distance in their catchment areas. The weed species also highly depend on seasons as maize is grown round the year in three seasons i.e. Kharif (Monsoon; June–September), Rabi (Winter; November–May) and Kharif 1 (Spring; February–June). The high weed seed pressure and diversity make more difficult to manage weeds manually in conservation agriculture as well as also in conventional agriculture. Grasses weed species during Kharif season in the maize field in EGP are *Echinochloa colona* (L.), *Digitaria ciliaris* (L.), *Leptochloa chinensis* (L.), *Dactyloctenium aegyptium, Cynodon dactylon, Echinochloa crus-galli, Eleusine indica, Setaria viridis, Panicum javanicum, Paspalum commersonii* etc. The broadleaf weeds are *Marsilea minuta, Polygonum hydropiper, Galinsoga ciliata, Physalis heterophylla, Heliotropium indicum, Phyllanthus niruri, Euphorbia hirta, Jussiaea repens, Amaranthus spinosus, Amaranthus viridis, Spilanthes paniculata, Lindernia anagallis, Paspalum distichum* etc. and the major sedges weeds are *Cyperus rotundus, Cyperus difformis, Cyperus iria, Eclipta prostrata, Ludwigia octovalvis, Portulaca oleracea, Fimbristylis miliacea, Scirpus* spp. etc. [29–31]. The perennial weeds species dominant round the years but the seasonal weeds are different when maize grown in dry season (winter maize) and the major dominant weeds are *Polygonum persicaria, Polygonum pensylvanicum, Polygonum orientale, Oldenlandia diffusa, Oldenlandia aquatic, Oxalis corniculate, Chenopodium arvensis, Physalis minima, Solanum nigrum, Hydrocotyle ranunculoides, Ageratum conyzoides, Medicago denticulate, Avena ludoviciana* etc. [32, 33].

A field experiment found that the most dominant weeds in maize field were sedge *C. rotundus* and dicot weeds *T. portulacastrum*, *D. arvensis*, *P. niruri* and grass *C. dactylon* [34]. In USA, it observed the most common broadleaves weed species in a maize crop were *Chenopodium album, Asclepias syriaca, Ambrosia artemisiifolia, Physalis heterophylla* and *Polygonum rensylvanicum* and the most common grasses were *Elymus repens and Setaria pumila* [35]. In another studies conducted on sandy loam soils at Hyderabad, India [26, 36], revealed that among the major weeds in maize field *E. colona* (grasses), *E. crus-galli, Paspalum distichum, C. rotundus* (sedges), *Ageratum conyzoides* and *T. portulacastrum, Sonchus oleraceus, Acalypha indica, Eclipta alba and Parthenium hysterophorus* (broad-leaved) were the predominant weeds. A 3 years field survey in Andhra Pradesh of India also confirmed that most dominant weed species in maize field were *E. colona*, followed by *P. repens*, *T. portulacastrum*, and *D. arvensis* [37].

4. The critical period of crop-weed competition

The critical crop growth stages considers as the most vulnerable period for crop-weed competition, during which crop must be weed free in order to prevent yield losses. Earlier studies observed [38–40] that the critical period of weed control in maize ranges from 7 to 56 days after seedling emergence. Other studies also reported [41–46], the critical period usually corresponds for maize up to 8–10 leaf stages. Wider canopy spacing and slow-growing nature of the maize crop should control weeds in first till 21–28 days after sowing for free from crop-weed competition and it was also suggested that if the weeds are not control within the critical crop growth stages, the yield losses may occur 30–100% [47, 48].

Weed species, densities, and their interactions influence maize yield loss [49, 50]. Weed plants compete with maize for their essential growth resources like water, nutrients, space etc. which ultimately reduce the yield up to 65% when weeds control measure was not performed at critical crop growth stages [46]. While, some problematic weeds species as they are similar in nature and life cycle of maize are difficult to control. Massinga et al., [51] reported that the yield reduction in maize could be 91% by competition if more than eight amaranth (*Amaranthus palmeri* S. Wats) plants per meter row length.

5. Weed control in zero-till maize by chemical measures

In maize production, weed management is considered as an important agronomic measure for attaining the potential yield. To minimize the maize yield loss due to weed competition, farmers are practicing several methods for controlling the weeds are available such as mechanical, cultural, biological and chemical control methods. The cultural methods are very expensive and time consuming so, farmers have to move towards other alternative methods of weed control [10]. Furthermore, due to the increasing cost and non-availability of labour for manual weeding during peak and critical maize growth stages significantly influence the maize yield. The role of herbicides is not only control the weeds timely and effectively, but also offer a great scope for minimizing the cost of production [10]. The chemical control method is quick, more effective, time and labour saving method than others [13]. However, it is important to use a broad-spectrum herbicide program including pre- and post-emergence herbicides for season-long effective weed control and to avoid shifts towards problematic weed species [32] or evolution of herbicide-resistant weed biotypes. On the other hand, it is decisive to select the appropriate weedicide depending upon the weed flora exist in a given field. In addition, the precise dose, methods, weed growth stage, timings, soil moisture and application techniques should be followed. A number of herbicides have been evaluated in sequential combination and suggested pre- and post-emergence herbicide application for effective weed control in dry direct drill-seeded rice systems, including under zero-tillage conditions [17]. Use of pre-sowing, pre- and post-emergence application of herbicides would make herbicidal weed control more acceptable to farmers which will not change the existing agronomic practices but will allow for complete control of weeds under CA based management practices. Pre-emergence herbicides spray will control the weeds up to 25 days after seeding and followed that post-emergence application depending on weed flora will take care to keep weeds below crop injury level. Pre-planting, pre and post-emergence herbicides either in sequential or tank mixture will be taken care of all types of weed flora an ideal means in view of economics and usefulness in maize [17, 31, 32].

6. Effect of pre-sowing herbicides on weed control of maize

Generally, pre-sowing/planting herbicides are non-selective which are applied to control prevailing complex annual and perennial weeds flora erstwhile to planting, particularly under the CA-based cropping system. Among them, herbicide Glyphosate (1 kg ai ha^{-1} or 0.5–1.5% by volume), Glucofosinate, or dicamba and Paraquat (0.5 kg ai ha^{-1} or 0.5% by volume) are the most widely applicable herbicides [52, 53]. Where Glyphosate and Paraquat herbicides are available and popular as these are systemic non-selective and contact herbicides, respectively, and kill both annual and perennial weeds. To be effective, the systemic (Glyphosate) should be applied when weeds are growing actively so that the herbicide is absorbed and translocated into the entire plant system [17], but contact herbicide (Paraquat) can be applied just before the sowing.

Currently, the present rates of many herbicides do not work properly against many weeds, due to their resistance against the target weeds. Therefore, it is important to apply a new product to be available earlier than normal in order to maximize the contribution of residual weed control after crop emergence. The earlier study found that if the pre-sowing herbicide is applied before planting and is incorporated in the soil with light tillage, the efficacy of the applied herbicides was found the maximum [54].

7. Effect of pre-emergence herbicides on weed control in maize

When herbicides apply immediately or 1–4 days after maize seeds sowing, and it can be also mixed with soil during sowing, but it must be applied before weed seed emergence, are known as pre-emergence herbicides. Earlier findings [55] showed that pendimethalin can be applied as pre-emergence to get maximum weed control efficiency and crop selectivity by decreasing the weed population and increased the maize grain yield over the weedy check field [56]. Thus, Mekky et al. [57] reported that when pre-emergence herbicides were applied immediately after seed sowing or pre-emergence, weed control efficiency was the maximum and also increased the maize yield.

A field experiment in the clay loam soils of Guntur, Andhra Pradesh (India), with zero tillage maize found that atrazine 1.5 kg ha^{-1} applied at pre-emergence followed by (*fb*) manual hand weeding (HW) at 30 DAS recorded the tallest plant and the maximum dry-weight over un-weeded check at all stages of crop growth [30]. Scientist [26] reported that when maize was grown under zero till condition with pre-emergence herbicides atrazine at 1.0 kg ha^{-1} and topramezone 0.030 kg ha^{-1} in combination of two hand weeding at 20 and 40 DAS was produced significantly higher plant height, dry matter production of maize.

Recently, a study conducted in a clay loam and sandy loam soils of Bangladesh as a weed management strategy in pre-emergence and post emergence combinations and results showed that the even application of both Pendimethalin and atrazine separately reduces the weed biomass, weed population at 30 days after seeding which further provided better crop yield [58]. In another study clay soils of Rajendranagar (Telangana, India) was found that the highest crop dry matter was recorded with two times HW (at 20 and 40 DAS), which was at par with pre-emergence atrazine at 1.0 kg ha^{-1} + paraquat 0.60 kg ha^{-1}, followed by oxyfluorfen 0.150 kg ha^{-1} + paraquat 0.60 kg ha^{-1} [59]. The sequential application of pre-emergence followed (fb) protected spray of non-selective herbicide (Atrazine as pre-emergence at 1.25 kg ha^{-1} *fb* Paraquat 0.6 kg ha^{-1} at 3 weeks after sowing (WAS) or Pendimethalin as pre-emergence at 1.5 kg ha^{-1} *fb* Paraquat 0.6 kg ha^{-1} at 3 WAS) produced the significantly higher yield than weedy check [60]. An experiment was conducted with application of atrazine or glyphosate herbicide alone and tank mix application of selective (atrazine) and non-selective (glyphosate) herbicides and found that the grain yield was 170 and 70% more when atrazine + glyphosate (5.25 kg ha^{-1}) were applied as tank mixture than weedy check and sole application of atrazine or glyphosate, respectively [26]. Similar to previous study, a field research in the sandy clay loam soils of Kampasagar (Telangana, India) reported that when atrazine 1.25 kg ha^{-1} + paraquat 0.75 kg ha^{-1} were applied as a pre-emergence in tank mixture produced the significantly maximum grains cob^{-1}, cob diameter and 100 grain weight than other herbicides [61]. Similar grain yields were recorded when applied atrazine alone as pre-emergence 1.25 kg ha^{-1} (6.7 kg ha^{-1}) and atrazine 1.25 kg ha^{-1} + glyphosate 0.5 kg ha^{-1} (7.0 kg ha^{-1}) as tank mixture of pre-emergence.

8. Effect of post-emergence herbicides on weed control in maize

Application of herbicides after the emergence of maize and weed are well-known as post-emergence herbicides. Generally, post-emergence herbicides spray/apply in standing crop targeting weeds canopy by using the sprayer equipment.

The most popular/well-known herbicides which have been found to be effective when applied as a post-emergence for effectively control of weeds in CA-based maize system are Atrazine, Tembotrione (Laudis), Halosulfuron methyl (Sempra), Tembotrione (Laudis) + Atrazine, Halosulfuron methyl (Sempra) + Atrazine [58]. Earlier findings [62] revealed that pre-emergence herbicides pendimethalin and atrazine reduced the grassy and broadleaves weed population to a significant extent and among the pre-emergence herbicides, later (atrazine) resulted in a higher reduction in grass weed population at the early stage (20 days after seeding) than former (pendimethalin). But further he suggested, the post-emergence herbicides, the mixture of tembotrione + atrazine was more effective in controlling all classes of weed flora at 40 and 60 DAS. Tembotrione alone also showed good control of grasses and broad-leaved weeds. Atrazine as pre-emergence followed by (fb) tembotrione + atrazine as post-emergence found best combination and this combination reduced the weed dry matter to the tune of 98.7 and 97.9% at 40 and 60 DAS, respectively which ultimately resulted in significantly higher grain yields (11.57 t ha^{-1}) with maximum net returns.

However, when a single herbicide is used for a long time for controlling the same weeds it may create resistance against the specific weed(s). Therefore, long term basis continuous use of the same herbicide should be avoided. So, it should be rotated with the use of multiple herbicides with a different mode of action to avoid/delay weed resistance against specific weeds.

Scientists found that application of two or more compatible herbicides use as a tank mixture are more effective to control broaden the spectrum of weeds including grasses, broadleaves, and sedges than a single application [11, 63]. In the USA, Atrazine has been used since 1958 in several million hectares of maize field due to its low cost, control of a broad spectrum of broadleaf weeds, flexible application timing, and also can be used as both pre- and post-emergence in combination with several other herbicides. As a result of long-term and continuous use, the atrazine was found in accumulation in food products, groundwater, and aquatic systems [64]. Therefore, we should be careful to know the residual effect of applied herbicides when using the as short term as well as long term.

9. Precautionary measures during application of herbicides in the crop field

Herbicides are chemical compounds used to control weed species but could also phytotoxic to crops and harmful to animals through entering direct or indirect in food chain. Therefore, herbicides should be carefully selected considering the toxicity, residual persistence in soil and water bodies as well as cropping systems. The herbicide residual persistence can be affected the succeeding crops in a crop rotations and also the runoff of rain water from crop fields to water bodies which may cause the lethal and hazardous to water organism and human beings [65]. However, non-target living organisms that come to direct contact with weedicides can be harmful or poisonous especially if they have carry the chemical of the toxic action [66]. Therefore, while working with herbicides poses an always chemical poisoning risk to exposed workers if precaution measures are not properly followed.

With the use of herbicides, there is a certain risk of intoxication to directly exposed environment, food product, workers, as well as applied crops, since it depends on numerous factors. Therefore, before application of any herbicide, its short and long term toxicity/risk factors should be considered. The risk of worker intoxication with pesticides and herbicides, depends on several factors as how they will be handled and used. It can be grouped under two major factors; the toxicity of the respective herbicide and how it was exposed under specific working condition [66]. Therefore, it is an important to must follow the safety and health management guidelines while working and using the herbicide at all steps as set out by the industrial hygiene [67]. The steps are the ability to antedate, identify, diagnose, evaluate, and minimize the risks in the workplace. Therefore, the Preventive safety measures of worker(s)/personnel such as psychological measures; administrative: legislation, standards, and procedures; and hygiene, cleaning, maintenance, and safety of the environment also important during selection and application of pesticides [68].

10. Conclusion and policy implication

From the above discussion of the chapter, it is confirmed that weeds are the major challenge in CA crop production systems, where almost 16 to 42% yield reduction is occurred due to weed infestation and one-third of the total cost of cultivation is spent on weeding. On an average of 13.1% of crop reproduced is actually lost in the farmers' fields even after adopting in traditional weed control. The zero/strip-tilled/permanent beds or till the soil with fresh beds based crops production system is an alternate option through mechanized precision planting within a single pass. Although, the CA-based crop management techniques will be faced the major concern of weed management initially. Therefore, proper weed management is considered one of the most important prerequisites in CA-based crop cultivation systems including maize to ensure high crop yield. High weed pressure in association with maize, increase to lower the economic returns and, in extreme cases complete failure of the crop. Hence, judicious weed management in CA system is a critical factor for securing and sustaining food security. While, number of repetitive tillage operations increase the cost of cultivation, fuel consumption and delays planting in two ways by repetitive tillage operations followed by manual sowing. After post seeding of maize, farmers are facing major challenges for weed management due to lack of pre-sowing, pre- and post-emergence herbicides.

Since the traditional weed management in maize systems after 30–35 days after seeding; generally, farmers cut the weeds with hand weeding which further consumed more labour or sometimes usually reluctant to control weed in the maize field. However, sometimes they weeded by hand, which proves uneconomical due to be increasing labour wages as well as lack of labours due to migrating from the villages to urban areas for better livelihood. The hand labour based weeding in many developing countries consumes up to half of the total labour demand. Therefore, CA-based new agronomic management practices may be advocated to overcome the above challenges. To address the weed management problems in CA-based maize production under no-till systems with different chemical weed control is a potential means for controlling weeds and more economical compared to hand weeding. Now some herbicides are available in the market for controlling weeds since these should be needed to validate for controlling weeds as well as to know their residual effect on the environment.

Conflicts of interest

The authors declare no conflicts of interest.

Disclaimer

We hereby declare that the book chapter does not have any material which has been accepted to publish any journal or publisher, and also has no copy of any material in previously published, except where due permission and reference is made in the text.

Author details

Akbar Hossain[1*], Mst. Tanjina Islam[2], Md. Shohidul Islam[2], Nurislam[2], Sharif Ahmed[3], Khokan Kumer Sarker[4] and Mahesh Kumar Gathala[5]

1 Bangladesh Wheat and Maize Research Institute (BWMRI), Dinajpur, Bangladesh

2 Department of Agronomy, Hajee Mohammad Danesh Science and Technology University, Bangladesh

3 Bangladesh Office, International Rice Research Institute, Dhaka, Bangladesh

4 Irrigation and Water Management Division, Bangladesh Agricultural Research Institute (BARI), Gazipur, Bangladesh

5 International Maize and Wheat Improvement Center (CIMMYT), Dhaka, Bangladesh

*Address all correspondence to: akbarhossainwrc@gmail.com
and tanjimar2003@yahoo.com

References

[1] IndexMundi. Corn: Area & Production by Country in 1000 MT. 2019. Available from: https://www.indexmundi.com/agriculture/?commodity=corn&graph=area-harvested [Accessed: 05 July 2019]

[2] Andersen PP, Pandya-Lorch R, Rosegrant MW. World food prospects: critical issues for the early twenty -first century. Food Policy Report Vision-2020: International Food Policy Research Institute Washington, D.C. 1999. Available from: https://cgspace.cgiar.org/bitstream/handle/10947/1622/world%20food.pdf?...1 [Accessed: 05 July 2019]

[3] Ofori E, Kyei BN. Agrometeorology and maize production. Chpt13c; 2004. Available from: www.agrometeorology/chpt13c [Accessed: 05 July 2019]

[4] Ali MY, Waddington SR, Timsina J, Hodson D, Dixon J. Maize-rice cropping systems in Bangladesh: Status and research needs. Journal of Agricultural Science and Technology. 2009;**3**(6):35-53

[5] Islam KS, Momin MA, Krupnik TJ, Tiwari TP, Rossi FJ, Gathala MK. Best Management Practices for Hybrid Maize Production. Published by International Maize and Wheat Improvement Center (CIMMYT) Bangladesh; 2013. p. 36

[6] Mahmoodi S, Ali R. Estimation of critical period for weed control in corn in Iran. Proceedings of World Academy of Science, Engineering and Technology. 2009;**37**:67-72

[7] Zimdahl RL. Weed-Crop Competition, A Review. Second ed. Ames, Iowa: Blackwell Publishing; 2004

[8] John RT, Michel AC. Subplots facilitate assessment of corn yield losses from weed competition in a long-term system experiment. Agronomy for Sustainable Development. 2010;**30**:445-453

[9] Ahmed S, Salim M, Chauhan BS. Effect of weed management and seed rate on crop growth under direct dry seeded rice systems in Bangladesh. PLoS One. 2014;**9**:e101919. DOI: 10.1371/journal.pone.0101919

[10] Varga P, Bcrcs I, Rcisinger P, Busak P. The influence of soil herbicides on weeds in maize. In: Proc. German Conf. Weed Biology and Weed Control, Germany. Vol. 17. 2011. pp. 641-646

[11] Ahmed S, Chauhan BS. Efficacy and phytotoxicity of different rates of Oxadiargyl and Pendimethalin in dry-seeded rice (*Oryza sativa* L.) in Bangladesh. Crop Protection. 2015;**72**:169-174. DOI: 10.1016/j.cropro.2015.03.021

[12] Chikoye D, Ellis-Jone J, Riches C, Kanyomeka L. Weed management in Africa: Experiences, challenges and opportunities. In: 16th International Plant Protection Congress. 2007. pp. 652-653

[13] Ahmed SE, Shams HM, El-Metwally IM, Shehata MN, El-Wakeel MA. Efficiency of some weed control treatments on growth, yield and its attributes of maize (*Zea mays* L.) plants and associated weeds. Mansoura University Journal of Agricultural Sciences. 2008;**33**(7):4777-4789

[14] Chikoye D, Manyong VM, Carsky RJ, Ekeleme F, Gbehounou G. Response of speargrass (*Imperatacylindrica* (L.) Raeusch) to cover crops integrated with handweeding and chemical control in maize and cassava. Crop Protection. 2002;**21**:145-156

[15] Chikoye D, Schulz S, Ekeleme F. Evaluation of integrated weed management practices for maize in the northern Guinea savanna of Nigeria. Crop Protection. 2004;**23**:895-900

[16] Chikoye D, Udensi UE, Fontem AL. Evaluation of a new formulation of atrazine and metolachlor mixture for weed control in maize in Nigeria. Crop Protection. 2005;**24**:1016-1020

[17] Kumar V, Ladha JK. Direct seeding of rice: Recent developments and future research needs. Advances in Agronomy. 2011;**111**:299-360

[18] Gathala MK, Kumar V, Sharma PC, Saharawat Y, Jat HS, Singh M, et al. Optimizing intensive cereal-based cropping systems addressing current and future drivers of agricultural change in the northwestern indo-Gangetic Plains of India. Agriculture, Ecosystems and Environment. 2013;**177**:85-97

[19] Kumar V, Jat HS, Sharma PC, Balwinder-Singh, Gathala MK, Malik RK, et al. Can productivity and profitability be enhanced in intensively managed cereal systems while reducing the environmental footprint of production? Assessing sustainable intensification options in the breadbasket of India. Agriculture, Ecosystems & Environment. 2018;**252**:132-147. DOI: 10.1016/j.agee.2017.10.006

[20] Tow P, Cooper I, Partridge I, Birch C. Rainfed Farming Systems. Netherlands: Springer; 2011

[21] Childs D, Jordan T, Ross M, Bauman T. Weed control in no tillage systems: Purdue University cooperative extension service. In: Conservational Tillage Series CT-2. West Lafayette: Purdue University; 2001

[22] Nalewaja J. Weeds and conservation agriculture. In: Conservation Agriculture. Dordrecht: Springer; 2003. pp. 201-210

[23] Chhokar RS, Sharma RK, Jat GR, Pundir AK, Gathala MK. Effect of tillage and herbicides on weeds and productivity of wheat under rice– Wheat growing system. Crop Protection. 2007;**26**:1689-1696

[24] Shrestha A. Conservation Tillage and Weed Management. Oakland, California, USA: University of California, Division of Agriculture and Natural Resources; 2006

[25] Mansouri AD, Bararpour MT, Babaeiangelodar N. Effect of tillage method and row spacing on growth and yield of soybean and weed management. Journal of Agricultural Sciences and Natural Resources. 1980;**Winter**:11. Eighth year. The fourth number

[26] Parameshwari YS. Influence of Rice Crop Establishemnt Methods and Weed Management Practices on Succeeding Zero-till Maize. Ph.D Thesis. Professor JayashankarTelangana State Agricultural University. Hyderabad. 2013

[27] Balota EL, Calegari A, Nakatani AS, Coyne MS. Benefits of winter cover crops and no-tillage for microbial parameters in a Brazilian Oxisol: A long-term study. Agriculture, Ecosystems and Environment. 2014;**197**:31-40

[28] Varvel GE, Wilhelm WW. No-tillage increases soil profile carbon and nitrogen under long-term rainfed cropping systems. Soil and Tillage Research. 2011;**114**:28-36

[29] Reddy MM, Padmaja B, Veeranna G, Reddy DVV. Bio-efficacy and economics of herbicide mixtures in zerotill maize (*Zea mays*) grown after rice (*Oryza sativa*). Indian Journal of Agronomy. 2012;**57**(3):255-258

[30] Rao AS, Ratnam M, Reddy TY. Weed management in zero-till down maize. Indian Journal of Weed Science. 2009;**41**(1 & 2):46-49

[31] Gathala M, Sudhir-Yadav, Mazid MA, Humphreys E, Sharif A, Krupnik T, et al. Guidelines for Dry Seeded Aman Rice (DSR) in Bangladesh. IFAD and CSISA joint publication. Intl. Rice Res. Inst. (IRRI) and Intl. Mexico: Maize and Wheat Improvement Center (CIMMYT); 2014. p. 32

[32] Singh AP, Bhullar MS, Yadav R, Chowdhury T. Weed management in zero-till wheat. Indian Journal of Weed Science. 2015;**47**(3):233-239

[33] Rai A, Mahata D, Lepcha E, Nandi K, Mukherjee PK. A review of weeds in maize. International Journal of Current Microbiology and Applied Sciences. 2018;7:2906-2922. DOI: 10.20546/ijcmas.2018.708.308

[34] Mukundam B, Srividya S, Raja V. Productivity and economics of rice-zero till maize as influenced by weed management practices in southern Telangana region of Andhra Pradesh. Indian Journal of Weed Science. 2011;**43**(3&4):163-168

[35] Pleasant J, Burt RF, Frisch JC. Integrating mechanical and chemical weed management in corn (*Zea mays*). Weed Technology. 1995;**8**(2):217-228 [Maize Abst. 1995; 11(3); 240]

[36] Singh S, Sheoran P. Studies on integrated weed management practices in rainfed maize under sub-montaneous conditions. Indian Journal of Dryland Agricultural Research and Development. 2008;**23**(2):6-9

[37] Kiran GGR, Rao AS. Survey of weed flora in zero till sown maize in Krishna zone of Andhra Pradesh. The Andhra Agricultural Journal. 2014;**61**(3):494-496

[38] Perry K, Evans MR, Jeffery LS. Competition between Johnsongrass (*Sorghum halepense*) and corn (*Zea mays*). Proceedings Southern Weed Science Society. 1983;**36**:345

[39] Vernon R, Parker JMH. Maize/weed competition experiments: Implications for tropical small-farm weed control research. Experimental Agriculture. 1983;**19**:341-347

[40] Ghosheh HZ, Holshouser DL, Chandler JM. The critical period of Johnson grass (*Sorghum halepense*) control in field corn (*Zea mays*). Weed Science. 1996;**44**:944-947

[41] Knezevic SZ, Evans SP, Blankenship EE, Van Acker RC, Lindquist JL. Critical period of weed control: The concept and data analysis. Weed Science. 2002;**50**:773-786

[42] Evans SP, Knezevic SZ, Shapiro C, Lindquist JL. Nitrogen level affects critical period for weed control in corn. Weed Science. 2003a;**51**:408-417

[43] Evans SP, Knezevic SZ, Shapiro C, Lindquist JL. Influence of nitrogen level and duration of weed interference on corn growth and development. Weed Science. 2003b;**51**:546-556

[44] Knezevic SZ, Evans SP, Mainz M. Yield penalty due to delayed weed control in corn and soybean. Crop Management Journal. 2003a;**2**. DOI: 10.1094/CM-2003-0219-01-RS

[45] Knezevic ZS, Evans SP, Mainz M. Row spacing influences critical time of weed removal in soybean. Weed Technology. 2003b;**17**:666-673. Available from: http://www.regional.org.au/au/asa/2004/poster/2/4/1/947_knezevicsz.htm [Accessed: 05 July 2019]

[46] Page ER, Cerrudo D, Westra P, Loux M, Smith K, Foresman C, et al. Why early season weed control is

important in maize? Weed Science. 2012;**60**:423-430

[47] Sandhu KS, Singh T, Singh S. Weed competition of maize (*Zea mays*) fields in Punjab. Indian Journal of Weed Science. 1999;**31**:18-24

[48] Knezevic SZ, Weise SF, Swanton CJ. Interference of redroot pigweed (*Amaranthus retroflexus*) in corn (*Zea mays*). Weed Science. 1994;**42**:568-573

[49] Fausey JC, Kekks JJ, Swinton SM, Renner KA. Giant foxtail interference in non-irrigated corn. Weed Science. 1997;**45**:256-260

[50] Scholes C, Clay SA, Brix-Davis K. Velvetleaf (*Abutilon theophrasti*) effect on corn (*Zea mays*) growth and yield in South Dakota. Weed Technology. 1995;**9**:665-668

[51] Massinga RA, Currie RS, Trooien TP. Water use and light interception under palmer amaranth and corn competition. Weed Science. 2003;**51**:523-531

[52] Gupta RK, Ladha JK, Singh S, Singh R, Jat ML, Saharawat Y, et al. Production technology for direct seeded rice. Rice Wheat Consortium Technical Bulletin. 2006;**8**:16

[53] Chauhan BS, Yadav A. Weed management approaches for dry-seeded rice in India: A review. Indian Journal of Weed Science. 2013;**45**(1):1-6

[54] Khaliq A, Matloob A, Ahmed N, Rasul F, Awan IU. Late POST-emergence chemical weed control in direct seeded fine rice. Journal of Animal and Plant Science. 2012;**22**:1101-1106

[55] Wilson RG, Sbatella GM. Integrating irrigation, tillage and herbicides for weed control in dry bean. Weed Technology. 2014;**28**(3): 479-485

[56] Fazal M, Ali K, Khan I, Khan HU, Anwar M. Efficacy of various herbicides against weeds and their impact on yield of maize. Pakistan Journal of Weed Science Research. 2009;**15**(2-3):191-198

[57] Mekky MS, Nassar ANM, Attalla SI. Effect of weed control treatments on weeds, growth, chlorophyll, crude protein and yield of maize (*Zea mays* L.). Egyptian Journal of Basic and Applied Sciences. 2002;**17**(6):219-240

[58] Gathala MK, Tiwari TP, Islam S, Maharjan S, Bruno G. Research Synthesis Report: Sustainable and Resilient Farming Systems Intensification in the Eastern Gangetic Plains (SRFSI). CSE/2011/077. CIMMYT-ACIAR publication. 2018. Available from: www.cimmyt.org

[59] Yakadri M, Rani PL, Prakash TR, Madhavi M, Mahesh N. Weed management in zero till-maize. Indian Journal of Weed Science. 2015;**47**(3):240-245. Available from: http://isws.org.in/IJWSn/File/2015_47_Issue-3_240-245.pdf [Accessed: 05 July 2019]

[60] Srividya S, Chandrasekhar K, Veeraraghavaiah R. Effect of tillage and herbicide use on weed management in maize (*Zea mays*. L). The Andhra Agricultural Journal. 2011;**58**(2):123-126

[61] Pasha ML, Bhadru D, Krishna L, Naik RBM. Evaluation of different herbicides in zero tillage. The Madras Agricultural Journal. 2012;**99**(7-9):471-472

[62] Mitra B, Bhattacharya PM, Ghosh A, Patra K, Chowdhury AK, Gathala MK. Herbicide options for effective weed management in zero-till maize. Indian Journal of Weed Science. 2019;**50**(2):137-141

[63] Ahmed S, Chauhan BS. Performance of different herbicides in dry-seeded rice in Bangladesh. The Scientific World Journal. 2014;**2014**:14. DOI: 10.1155/2014/729418

[64] Ying GG, Kookana RS, Mallavarpu M. Release and behavior of Triazine residues in stabilized contaminated soils. Environmental Pollution. 2005;**134**:71-77

[65] Larini L, Cecchini R. A intoxicação como fenômeno biológico. In: Larini L, editor. Toxicologia. São Paulo: Editora Manole Ltda; 1987. pp. 1-40

[66] Bonsall JL. Measurement of occupational exposure to pesticide. In: Turnbull GI, editor. Occupational Hazards of Pesticide Use. London: Taylor e Francis; 1985. pp. 13-33

[67] Brazil. NR 9—Programa de prevenção de riscos ambientais. Brasília, 2014. Available from: http://portal.mte.gov.br/data/files/FF80808148EC2E5E014961B76D3533A2/NR-09%20(atualizada%202014)%20II.pdf [Accessed: 05 July 2019]

[68] Machado-Neto JG. Safety measures for handlers/workers against herbicide intoxication risk. In: Price A, Kelton J, Sarunaite L, editors. Herbicides, Physiology of Action, and Safety. United kingdom: Intech Open; 2015. Available from: https://www.intechopen.com/books/herbicides-physiology-of-action-and-safety/safety-measures-for-handlers-workers-against-herbicide-intoxication-risk [Accessed: 05 July 2019]

Chapter 9

Women Participation in Post-Harvest Processing of Maize Using Indigenous Technologies: A Perspective of Kogi State of Nigeria

Adejo Patrick Emmanuel

Abstract

This research was carried out in Kogi State, Nigeria. This study investigated the level of women participation in post-harvest management practices of maize in Kogi State. Women participation in post-harvest handling of crops in Nigeria and particularly in Kogi State is a focal area for researches in the post-harvest subsector of agriculture. Women are found to be actively engaged in making sure that harvested crops are properly stored and processed into forms that are consumable and marketable. Data collected for this study were analysed using descriptive statistics and Chi-square analysis. Results indicated greater percentages (42.26 and 41.67%) of women farmers highly participated in the storage and marketing of maize, but low participation in transportation, grading/sorting/packaging and processing of maize was recorded. it is recommended that women should have access to productive resources (land, labour and credits, technical and economic information) in order to boost their participation in post-harvest management and value addition of maize. This chapter deals with women activities in indigenous post-harvest management of harvested maize which covers storage, processing, sorting, grading, packaging, transportation and marketing. It analyses the peculiarity of women engagement in post-harvest activities in Nigeria, and particularly in Kogi State.

Keywords: maize, women, post-harvest, indigenous, participation

1. Introduction

Agricultural activities are characterized by gender division of labour in the Sub-Saharan Africa and in Nigeria in particular. Thus, women are mostly found to be engaged in post-harvest activities than their men counterpart whose activities revolve round pre-production and actual production operations that are neck twisting and backbreaking, even when women do, their participation in these operations are minimal and largely dependent on the socio-cultural and agrarian setting of the place in question. The Womenfolk plays a key role in agriculture, especially in the activities revolving round harvest and postharvest. However, their contribution is always ignored and relegated to the background. It is reported by [1] that in rural

Pakistan, women carryout 53% of farming activities, and devote 20% more time than men for work. Yet, they face more constraints than men as they are negatively affected by socio-cultural values, economics policies and decision making in the society. Lately, studies have shown that the underlying reasons of food losses are associated with specific socio-cultural and gender dimensions [2, 3].

Maize is a major staple crop in Nigeria and across Africa. It is said to be the second most common cereal food crop after rice [4]. Green fresh maize is cooked or roasted and hawked by women and children, providing a livelihood for many urban poor households. It is also used for animal feeds and in industries such as flour mills, breweries and confectioneries. Thus, increased maize production will enhance food security, serve as import substitution and earn foreign exchange for the country through export to neighbouring food deficient countries and potentially beyond [1]. In most places across the country, women are found to mostly undertake activities that range from harvesting, transportation of harvested crops from the farms to homes, processing, preservation, packaging and marketing [5]. The women equally provide about 60% of agricultural labour force and contribute to well-being of their households through their income generating activities [6].

Historical trends have suggested that the majority of increased maize production in Nigeria reflects an expansion in land under its cultivation rather than an improvement in yields [7]. In 2010, the estimate for Corn production in Nigeria was put to about 8800 Metric tonnes with growth rate of −1.68% and in the year 2012, it rose to about 9410 Metric tonnes for which the growth rate was put to 1.73%. In recent time, Nigeria's yield per hectare is found to be lower than the global yields in the 1960s (1.9 tons/ha). It falls in the range of 1.63 and 1.76 tons per hectare between 2004 and 2007, this is lower than the global average which could be found in the range of 4.88 and 4.93 tons per hectares for the same period [8]. Meanwhile, Kogi State recorded about 333.21 Metric tonnes of corn production with the area under cultivation of 206.60 Hectare in 2009 and in 2010, it increased minimally to 371.30 MT with area under cultivation of 227.05 hectares [9]. Kogi State is one of the 36 states in Nigeria and it is in the North-central of the country.

Maize post-harvest losses in the tropics have been estimated to be about 20% [10]. Estimated losses of Nigeria maize post-harvest to have ranged from 15 to 20%. Nigeria has a land area of 98.3 million hectares and at present about 34 million hectares or 48% are under cultivation. It has been found that more than 60% of maize produced in Nigeria's is used industrially for the production of flour, beer, malt drink, corn flakes, starch, syrup, dextrose and animal feeds. An embargo has been placed on the importation of maize in order to boost local production to meet the consumption demand of Nigeria. This reveals significant room for the improvement of maize production in Nigeria which is necessary if the newly developing trend of exportable surplus is to be sustained and expanded.

According to [11, 12] the post-harvest operations of maize include drying, shelling, threshing, storage and marketing. Apart from the foregoing activities [13], added grading and sorting, packing and bagging, transportation, loading and unloading. Post-harvest management of maize can be categorized into four major stages [5]. The first stage is farm-gate operations which include de-sheathing and packing harvested maize cobs together when they are mature.; the second is transportation, when cobs are transported from farms to homesteads; the third is homestead activities which constitutes drying and grading of cobs, shelling of cobs and drying the grains, winnowing of grains, application of pesticides, packing grains at household level and the fourth stage is storage of maize where grains are stored for later use. Studies have shown that proper post-harvest handling can improve earnings of maize growers especially for the women folk [14, 15]. This study focused on assessing the level of women participations in post-harvest management on maize in Kogi State of Nigeria.

2. Methodology of the study

This study was carried out in Kogi State, Nigeria. The State consists of 21 Local Government Areas. The State is located between latitude 6° 30′N and 8° 5′N and longitude 5° 51E and 8° 00E. The state is made up of four agro-ecological zones, namely: Zone A with the headquarter in Aiyetoro Gbede; Zone B has its headquarter in Anyigba; Zone C with headquarter in Koton Karfe; and Zone D has its headquarter in Alloma. This delineation or stratification of the State into these Zones was done By Kogi State Agricultural Development Projects (KSADP) in order to ease their extension service delivery for optimum results in the State.

The state is well endowed with river valleys and swamps lands for dry season farming. Maize is one of the major crops grown in the state and thrives well in virtually all the agro-ecological zones of the State. For the purpose of this study registered farmers were randomly selected. A total of 168 farmers were selected for the study. A three stage random sampling technique was used. In stage one, four (3) extension blocks were randomly selected from each of the agricultural zones (A, B, C, D), that gives a total of 12 extension blocks. There are five blocks in each zone. In stage two, three (2) extension cells were randomly selected from each block, making a total of 24 extension cells. There are 8 cells in each zone. In stage three, five (7) registered farmers were randomly selected from each cell. There are 80 farmers in each cell. A total of 168 farmers were used for study. Structured questionnaire and interview scheduled were used to elicit information from the respondents. The date collected was analysed using both descriptive and inferential statistical tools. The hypothesis for the study was tested using Chi-square (X^2) statistics. The calculation of Chi-square statistics is given as follows:

$$X^2 = \frac{\Sigma(fo\text{–}fe)2}{fe} \quad (1)$$

where χ^2 = Chi-square value.

f_o = the observed frequency of levels of women participation in post-harvest operations in the Zones (A, B, C and D).

f_e = the expected frequency of levels of women participation in post-harvest operations in the Zones (A, B, C and D).

df = (R-1) (C-1).

Decision rule: the Null Hypothesis is rejected when $\chi 2$ Computed ≥ $\chi 2$ Tabulated under the degree of freedom.

Null hypothesis (Ho): there is no significant difference in the levels of post-harvest operations of maize in the various agro-ecological zones (A, B, C and D) of the study area.

3. Results and discussion

3.1 Indigenous post-harvest handling of maize by women farmers

The post-harvest management of maize can be discussed under the following subheadings: Storage, processing grading/sorting/packaging, transportation/loading and unloading and marketing of Maize as contained in the **Table 1**.

3.1.1 Storage of maize

The analysis of level of women participation in the post-harvest operations as presented in **Table 1** shows that 16.07, 36.91, 42.26 and 4.76% of the women maize farmers rated their participation in storage of maize as low, moderate, high and no participation respectively. Drying and storage of maize grains have always been considered as part of the domestic activities normally performed by women. This is an indication that women farmers in Kogi State participated highly in the post-harvest operations of maize. This result is in consonance with that of [16] who reported that 42.2% of women actively engaged in drying and storage of maize in five agro-climatic zones of rural Punjab, India. The construction of the storage facilities are mostly done by men. Although, women sometimes hire the men and youth in helping them construct the crib as shown in **Figure 1**. This construction of crib in the study area is in line with the report of [2] that carried out a study on gender post-harvest activities in Zambia, that the male support their female counterpart in constructing the storage facilities.

Effectiveness of the preservation of maize, like that of other cereal grains and food legume, is largely hinged on the ecological conditions of storage; the physical, chemical and biological features of the grain; the period of storage; and the type and the features of the storage facilities. Maize can be stored for a long period of time in the raw form. Its shelf life is largely anchored on the prevailing weather conditions (temperature and humidity) and other factors like the moisture, pests in the stored maize and diseases. Hence, it is worthy of note that proper post-harvest management of maize involve consideration of the above factors in order to improve the shelf life of maize. The recommended quality of maize is highly dependent on control harvesting strategies employed.

Maize harvesting is done by separating the cob from its main stalk. The appropriate period of harvesting maize is when the stalk is dried and moisture of grain as about 20–17% [17]. Maize that is harvested dry contains 15–20% moisture at the time of harvesting. After harvesting, the husks are removed from the cob. Then shelling is done to separate the grants from the rest of the cob. In many village settings, the shelling is done by hand, or by rubbing the cob over a roughened piece of metal. However, the use of mechanical shellers for maize is fairly wide spread in West Africa, and adopting the mechanical method for shelling their maize.

Post-harvest operations	Low (1) Freq.	Moderate (2) Freq.	High (3) Freq.	No participation (0)
Storage	27 (16.07)	62 (36.91)	71 (42.26)	8 (4.76)
Processing	60 (35.71)	47 (27.98)	28 (16.67)	33 (19.64)
Grading/sorting/packaging	54 (32.14)	29 (17.26)	39 (23.21)	46 (27.38)
Transportation/loading and unloading	67 (39.88)	45 (26.79)	34 (20.24)	22 (13.10)
Marketing	32 (19.05)	46 (27.38)	70 (41.67)	20 (11.91)

Field Survey, 2014.
Note: multiple responses.
The figures in parenthesis represent percentages (%).

Table 1.
Distribution of maize farmers according to the level of women participation in the post-harvest operations of maize (n = 165).

Figure 1.
Maize cobs stored in a typical crib in Bassa, Kogi State.

In order to reduce the moisture content of maize, it has to be dried to 10–11% moisture content before storing it [18]. Some have advocated drying it to moisture of 13–14% [19]. However, the former seem to increase the shelf life of stored maize because the dryer the maize, the better it is stored. For proper drying of maize, it may be left spread on a mat or any platform over several days; it could also be taken to the drying machine which blows hot air over the grains. After it must have been dried to the desired moisture content, the maize grains can then be stored in silos (metal or earthened) or in bags over a long period of time without deterioration. However, if the grains have not been thoroughly dried, they will deteriorate and become mouldy during storage.

In many rural parts of West Africa, maize is not shelled before being stored. Instead it is stored on the cob, with or without the husk [7]. Quite frequently, the maize is hung over domestic fireplaces where the fire helps make it dry, keep it dry and discourage weevil attack. More commonly, the maize is stored in large circular cribs or bins, built of bamboo and palm material, on raised platforms. Fires are occasionally built underneath the bins to promote drying and to control insect pests in the stored cobs. In these methods, shelling is done only just before the maize is to be utilized or taken to market. Factors that governed the design of storage structures included: tribal inheritance, availability of local building materials, social, economic and cultural standards of the people and local customs.

Even though farmers have adopted some innovations in post-harvest management of their crops, farmers continue to be glued to their indigenous post-production practices because of their effectiveness and reliability over time. For example, farmers in the eastern part of Kogi State do not use industrial dryer on the corn that are meant for the next planting season as they believe it will destroy the viability of planting. Instead, they prefer the indigenous way of storing it on cobs local silos or cribs, basket or hung near fire [14]. Effective management during the post-harvest period, rather than the level of sophistication of any given technology, is the key in reaching the desired objectives.

3.1.2 Processing of maize

Spoilage of harvested maize is prevented wholly or partly either by appropriate storage or by processing it into various storable products. Maize is consumed in many forms in different parts of the world, from maize grits, maize porridge, polenta and corn bread to popcorn and products such as maize flakes [11]. With respect to processing of maize as indicated in **Table 1**, it was found that 35.71, 27.98, 16.67 and 19.64% of the women maize farmers said their participation in processing of maize into consumable and marketable forms was low, moderate, high and no participation respectively. This result is in agreement with a study by [2, 16] that processing activities was low among women in Punjab, India and Malawi. Processing activities in this study is associated with those of both primary and secondary transformation of maize into consumable forms. This primary processing includes threshing, winnowing, cleaning, soaking, dehulling, grinding, while the secondary processing are activities involves changing maize into forms ready for the table. They are activities such as boiling, steaming, roasting, frying, crushing, blending, cooking and baking. Primary processing of maize is very essential in the study area as maize flour is highly demanded for across all the agro-ecological zones of the State.

Traditional methods of processing maize into flour using pestle and mortar, grinding stones and manually operated mills vary in detail depending on culture and geography [10]. The principles however, are the same in Kogi State, Nigeria. Maize processing essentially begins with soaking the grain and then grinding it between stones or pounding it in a mortar and pestle arrangement. During pounding or grinding the bran is removed. The grain is winnowed at intervals to remove the bran from the kernels. The dehulled grain is then pulverized into flour by further grinding or pounding. Processing of maize into desirable end products usually involves primary processing (cleaning, grading, soaking dehulling, grinding and sieving) and secondary processing (blending, cooking, frying and baking).

In most West African countries, maize is commonly used to make local beer (*Burukutu*). For example, malt is gotten from maize grains left to germinate for 4–5 days [10]. The malt is then exposed to the sun to stop germination. The grains are lightly crushed in a mortar or on a grinding stone. The malt is cooked and the extract is strained off, cooled and allowed to stand, after 3 days of fermentation it is ready to be drunk as beer. In some part of Nigeria like in the East of Kogi State, millet is mostly mixed with maize for the above process of brewing except that yeast is added and left for some few hours instead of fermentation for 3 days. Threshing is mostly done by farmers beating the maize cobs in mortar using a pestle or beating the cobs with a stick while they are still spread on the floor or rocky areas covered with sack. Using improved method of threshing, the maize in its cobs could be conveyed to the shelling machine (Thresher/Sheller) of which very few of them were available at the time of this study or further processing including dehulling and grinding machine. **Figure 2** depict a typical mortar and pestle for threshing maize, while **Figure 3** shows women dehulling and grinding maize grains. Winnowing and sun drying can be carried out before dehulling and grinding. **Figure 4** illustrates woman winnowing and sun drying maize.

Cornflakes are a hydrothermically treated maize product of world-wide popularity. They owe their success to their high nutrient value combined with low caloric content and good digestibility. Flaking is a process consisting of cooking fragments of cereal kernel, to a certain consistency, pressing the cooked mass between rollers to form flakes, and toasting the flakes at an appropriate temperature.

Figure 2.
Mortar and pestle for threshing maize grains from the cobs.

Figure 3.
Women dehulling and grinding maize grains.

3.1.3 Sorting, grading and packaging of maize

In terms of sorting, grading and packaging of maize in the study area as shown in **Table 1**, 32.14, 17.26, 23.21 and 27.38% rated their participation to be low, moderate, high and no participation respectively. This result indicates that the participation of women farmers in Kogi State in the grading/sorting/packaging of maize was rated low. Sorting and grading are done during threshing of maize whereby, maize grains are winnowed with the lighter grains separated in as it is thrown from the container into the air taking into cognizance the direction of the wind; and bad grains can also be separated from the good ones by hand picking the bad grains as shown in **Figure 3**. However, packaging is done by putting the grains into woollen or polythene bags, while labelling is done by using chalk or ink inscribed on the bag with any sign or words for identification. Indigenous

Figure 4.
A woman winnowing and sun drying maize.

value added products like maize porridge (*ekwo, ogidigbo agidi*) are packaged in banana leaves or other broad leaves that can be folded around the semi-solid porridge. The proportion of the respondents that did not respond may be due to fact that they are used to their indigenous way of grading/sorting/packaging of maize and too, that the improved grading and packaging available may not be cost effective for the farmers. For instance, the improved bagging system (triple bags), at the time of this report were sold at N300 per triple bag as against the N50 per single ones.

3.1.4 Transportation of maize by women farmers

With respect to transportation of maize, 39.88, 26.79, 20.24 and 13.10% of the women farmers said their participation was low, medium, high and no response respectively. This means that the participation of women farmers in Kogi State in the transportation of maize was rated low. Transportation is done indigenously by using their head with a basket or any other container to carry the maize cobs or grains. The load could be transported using wheelbarrow or bicycle to convey harvested maize from farm to home or markets or from stores at home to the market. It is noteworthy, that women are sometimes helped by their men counterpart especially those that are married in terms of transportation of maize particularly in the area of loading and unloading, if the harvested maize is bulky. However, in some cases, improved transportation, such as lorry, pick-up van, trucks, etc., could be employed for bulkier loads especially at farm gate where many small-scale farmers put their produce together for buyers (bulk assemblers) (**Figure 5**).

3.1.5 Maize marketing by women farmers

Maize like most agricultural commodities can be marketed freshly harvested or processed. In Kogi State, local market places which are normally held at an interval of 4–5 days are the points of convergence of the farm produce, including maize. Maize processing would further enhance the chances of success in its marketing. In terms of marketing of maize in study area as reflected in **Table 1**,

Figure 5.
Wheelbarrow for conveying load.

Figure 6.
The researcher observing the triple bagging system in front of a warehouse in the market.

19.05, 27.38, 41.76 and 11.91% of the women maize farmers rated their participation as low, moderate, high and no participation respectively. Women in Kogi State were found to be mostly responsible for price taking and giving in maize grain marketing. This does not mean that men do not take part in marketing of maize; but their activities in marketing of maize are minimal when compared to women involvement in same. One important feature of the markets in the study area is the presence of warehouses in and around the market which makes it easier for marketers to store their goods. **Figure 6** depicts a typical market with warehouses.

One problem faced by both small and large scale maize farmers in food processing particularly if they are into traditional foods, is how to market their products at a price that will guarantee a reasonable margin of profit. The chances of success of a large scale commercial venture producing traditional food could be further enhanced, especially if they are part of a group or chain of industries, and if some of the materials which they plan to produce will be utilized by one of the other arms of the industrial establishment.

Post-harvest operations	Chi-square (X^2) value	P-value
Storage	11.22^{*}	0.10
Processing	8.99^{NS}	0.10
Grading/sorting/packaging	0.85^{NS}	0.10
Transportation/loading and unloading	13.29^{*}	0.05
Marketing	16.70^{*}	0.01

*Significant.
NS Not significant.

Table 3.
Chi-square analysis on the level of women participation in post-harvest operations of maize in the various zones (A, B, C and D) of Kogi State, Nigeria.

Level of participation in storage of maize			
Zones	**Low**	**Medium**	**High**
A	8	15	18
B	3	25	30
C	10	10	12
D	6	12	11
Level of participation in processing of maize			
Zones	**Low**	**Medium**	**High**
A	10	12	6
B	20	18	10
C	20	11	4
D	10	6	4
Level of participation in grading/sorting/packaging of maize			
Zones	**Low**	**Medium**	**High**
A	13	8	10
B	18	8	10
C	10	6	9
D	13	10	10
Level of participation in transportation of maize			
Zones	**Low**	**Medium**	**High**
A	14	11	4
B	15	22	12
C	20	6	8
D	18	6	10
Level of participation in marketing of maize			
Zones	**Low**	**Medium**	**High**
A	6	10	20
B	6	18	31
C	10	4	6
D	10	14	13

Table 2.
Frequency distribution of respondents according to their level of participation in post-harvest operations in the various zones of Kogi State.

3.2 Tests for hypothesis on the level of women participation in post-harvest operations in the various zones (A, B, C and D) of Kogi State

Table 3 shows the analysis of Chi-square on the level of post-harvest activities among farmers in the various zones (A, B, C and D respectively). This was achieved using the observed frequencies in **Table 2**. The result indicated that there was a significant difference ($P \leq 0.10$, $P \leq 0.05$ and $P \leq 0.01$ respectively) their level of post-harvest operations in storage, transportation and marketing. This implies that the null hypothesis which was stated that 'there was no significant difference in the levels of post-harvest operations of maize in the various agro-ecological zones (A, B, C and D) of the study area' was rejected in each case. This may not be unconnected with the fact that maize thrives well in all the agro-ecological zones of the State and by extension, a high participation of women in the post-harvest activities should be expected across the zones of the state all things being equal. There were no significant differences in in terms of grading/sorting/packaging and transportation. This could be tied to the fact that these marketing functions are activities that relatively depend on the presence of designated market places; hence, these post-harvest operations may be higher in the areas where these markets were found, and lower if otherwise.

4. Conclusion

It can be inferred that there is generally, a high participation of women farmers in the storage of maize in the study area, but low participation in transportation, grading/sorting/packaging and processing of maize were recorded. This could be due to some factors that might have directly or indirectly affected the participation of respondents. The differences that occurred in the level of participation by women in post-harvest management of maize at the various agro-ecological zones of the State because of the peculiarity of the dominance of some post-harvest handling operation in a particular zone. This can be seen as a point to harness and integrate the indigenous knowledge on post-harvest handling of maize from these areas in order to boost their post-harvest activities. Their low level of participation in some of the post-harvest operations could be tied to the fact that they had inadequate access to information and other productive resources. If they have access to productive resources, they could improve their level of participation in post-harvest activities. The adequate knowledge and attitude about the appropriate indigenous post-harvest technology to be used by the farmers for maize in a particular agro-ecology will go a long way to boost farmers' participation in post-harvest management of maize. It is obvious that despite the existence of various improved post-harvest technologies in Nigeria, most of the women in the study area are still glued to some indigenous technologies for the reason that they might have tested and are familiar with these indigenous technologies and perhaps, they have little or no access to the improved or modern ones.

Author details

Adejo Patrick Emmanuel
Faculty of Agriculture, Department of Agricultural Economics and Extension, Kogi State University, Anyigba, Kogi State, Nigeria

*Address all correspondence to: adejo.pe@ksu.edu.ng

References

[1] Humera A, Tanvir A, Munir A, Muhammed IZ. Participation level of rural women regarding post-harvesting activities in Pakistan. Pakistan Journal of Life and Social Science. 2009;7(2):136-139

[2] Dyton OM, Vongai K. Gender analysis of maize post-harvest Management in Malawi: A case of Lilongwe and MChinji districts. A Report Submitted to Swiss Agency for Development and Cooperation (SDC), Swiss Cooperation Office Southern Africa; 2015. pp 33-55

[3] Sawadogo-Ouedraogo H, Diancoumba D, Tagnan A, Totobesola M, Safa A. Gender dimension in the post-harvest losses analysis and recommended solutions in Burkina Faso: Sorghum, maize and cowpea. In: 1st Africa Post-Harvest Congress and Exhibition; 28-31 March 2017; Safari Park Hotel, Nairobi, Kenya; 2017

[4] Ajani O, Okoruwa VO, Rahji AM. Rice and Maize Stratification Project in Nigeria: Draft Report to the Federal Ministry of Agriculture. 2007

[5] Okwoche VA, Age AI, Akor ID. Indigenous grains storage facilities in Ankpa local government area, Kogi state, Nigeria. International Journal of Agricultural Management and Development. 2010;**1**(1):101-107

[6] Adisa BO, Okunade EO. Women in agriculture and rural development. In: Adedoyin SF, editor. Agricultural Extension in Nigeria. Agricultural Extension Society of Nigeria (AESON), Ilorin, Kwara State, Nigeria; 2005. pp. 69-77

[7] Ogbona KI, Okoroafor E. Enhancing the capacity of women for increased participation in Nigeria main-streaming agriculture: A re-designing of strategies. Journal of Farm Management Association of Nigeria. 2004;7(2):33-48

[8] USDA. 2012. Nigeria Corn Production by Year. Available from: www.chicagotribune.com [Retreived: 27 June 2012]

[9] KSADP: Kogi Agricultural Development Project. Annual Report by Planning Monitoring and Evaluation Department of KSADP; 2011. p. 15

[10] Kalunda P, Agona A, Okot CT. Improvement of maize marketing through adoption of improved post-harvest technologies and farmers group storage. A case study of Kiboga and Apac districts of Uganda. Research for Development. 2004

[11] Adebayo OO, Oyetoro JO. Perception of agricultural information needs by small-scale maize farmers in Isin local government area of Kwara state, Nigeria. The International Journal of Agricultural Economics, Extension and Rural Development. 2011;**4**(1):56-61

[12] Mtambanengwe F, Kosina P, Jone J. Maize Seed Management-Post-Harvest Treatment and Storage. Cereal Knowledge Bank. 2009. Available from: http://www.knowledgebank.irri.org/conta ctuUs.htm [Accessed: 14 January 2013]

[13] World Bank. 2011. Missing Food: The Case of Post-Harvest Grain Losses in Sub-Saharan African. World Bank Report No: 60371-AFR. Available from: www.worldbank.org/rural [Retrieved: 14 November 2014]

[14] Monde MM, Vongai K. Gender analysis of maize post-harvest management in Zambia. A case study of Chipta and Katete districts. A Report Submitted to Swiss Agency for Development and Cooperation (SDC),

Swiss Cooperation Office Southern Africa; 2015. pp. 2-3

[15] The New Times. How Proper Post-Harvest Handling, Storage have Improved Earnings of Gisagara Maize Growers. The New Times. 2018. Available from: https://www.newtimes.co.nw/category/2906/business

[16] Sidhu K. Participation pattern of farm women in post-harvesting. Studies on Home and Community Science. 2007;**1**(1):45-49

[17] Ethiopian Agricultural Research Organization (EARO 2000). Food Science and Post- Harvest Technology Research Strategy. Nazareth, Ethiopia; 2000. pp. 9-14

[18] Aseidu EA, Van Gastel AJG, Greg B, Ebert A. Dehumidifying: Available option for long term seed storage in the humid tropics. In: Proceedings of a Regional Maize Workshop; 4-7 May 1999; IITA—Cotonou, Benin Republic; 1999. pp. 59-67

[19] Daily Trust. Guideline for Post-Harvest Handling of Maize. 2018. Available from: https://www.dailytrust.com.ng